T0318775

Exploring Mathematical Modeling in Biology Through Case Studies and Experimental Activities

Exploring Mathematical Modeling in Biology Through Case Studies and Experimental Activities

Rebecca Sanft
University of North Carolina Asheville
Asheville, NC, United States

Anne Walter
St. Olaf College
Northfield, MN, United States

Academic Press is an imprint of Elsevier
125 London Wall, London EC2Y 5AS, United Kingdom
525 B Street, Suite 1650, San Diego, CA 92101, United States
50 Hampshire Street, 5th Floor, Cambridge, MA 02139, United States
The Boulevard, Langford Lane, Kidlington, Oxford OX5 1GB, United Kingdom

Library of Congress Cataloging-in-Publication Data
A catalog record for this book is available from the Library of Congress

British Library Cataloguing-in-Publication Data
A catalogue record for this book is available from the British Library

ISBN: 978-0-12-819595-6

For information on all Academic Press publications
visit our website at https://www.elsevier.com/books-and-journals

Publisher: Mara Conner
Acquisitions Editor: Chris Katsaropoulos
Editorial Project Manager: Emma Hayes
Production Project Manager: Poulouse Joseph
Designer: Mark Rogers

Typeset by VTeX

Cover credit: Cover photo taken by Michael C. Swift and modified by Anne Walter. Orchid plant in the cloud forests of Ecuador.

Contents

Preface

This mathematics of biology text is a stand-alone compendium of exercises, cases, and wet labs designed to help mathematics and life sciences college students integrate mathematical, computational, and research approaches to addressing real problems. These materials are designed for any course in the broad area of mathematical biology for undergraduates, particularly filling a niche for classes intended for students from multiple disciplines. There is an ongoing paradigm shift in how science is practiced and taught, from a strong disciplinary focus to one of interdisciplinary approaches to address a common problem. One critical interface is that between mathematics and biology, which is exploding with possibilities due to the emergence of powerful computational and data collection tools allowing models and data to converge in hypothesis testing and evaluation.

This compendium provides a set of materials to actively engage students in mathematical modeling of biological phenomena while simultaneously interacting with real data gathered from the literature or experimentally by them. We intend students to experience how the act of modeling brings clarity to complex systems and often identifies or helps scientists extract parameters that are most critical or unavailable experimentally. Students will see the utility of models for identifying further questions, predicting outcomes (drug dosing plans, invasive species population growth, or effects of global warming) or understanding complex systems (community population dynamics, feedback loops, and hormonal control).

This text is truly interdisciplinary. It provides case studies across multiple levels of organization and areas of biological inquiry and includes instructions for several wet labs so that students can collect their own data to use in model formulation and validation. Each unit has specific learning goals, mathematical background, embedded R code, and examples and exercises in the form of case studies and a lab, thus making it easy to adapt some parts or all of this text for a wide range of courses that integrate mathematics and biology from the level of several homework assignments or a case study up to an entire course. An example of the latter based on our experience is given in [61].

The audience for this compendium of lessons, exercises, case studies, and lab exercises is primarily undergraduate life sciences or mathematics students and their faculty. It is designed as a first course in mathematical modeling with a focus on biological problems. Although the target audience is sophomores, we have had students at all undergraduate levels successfully learn from and enjoy these materials. We do assume the reader has had single-variable calculus but no college-level biology; the labs are typical of those in nonmajors courses with significantly higher expectations for data analysis. Due to the latter requirement, advanced biology students benefit tremendously from looking at an experiment from a different point of view.

This text is designed to foster students' abilities to:

- Clearly state objectives of a model in terms of biological and mathematical outcomes.
- Translate objectives and current knowledge of the system into specific hypotheses.
- Formulate an appropriate model, define the meaning and units of all variables and parameters, and articulate model assumptions.
- Estimate model parameters from experimental data.
- Interpret model output and data in terms of the original biological question.
- Use results to ask follow-up questions or suggest experiments.

Tips for using this text – instructors

How you use this text is up to you, but it might be helpful to know how each unit is organized to help you decide what works best. There is a lot of material for a one-semester course, so we expect that most users will select items that work best for them. For example, if the emphasis of your course is ecological, then you might choose only the case studies with that focus and skip the more organismal or medically driven examples. We also expect that some of the examples, cases, and labs might be used in other types of courses; these can be used on their own, but it may help to refer to some background information and R code in the text.

We start with a short section on R Basics to introduce the coding environment and practice using this tool for calculations, importing data, plotting, iteration, and performing linear regression. This section is very handy if you are new to R and a good reference for anyone using R. Additional R commands are introduced as needed throughout the text.

We have included a "Prelab Lab" intended for students with little or no lab experience. We have found that it is a good way to ease novices into the lab and simultaneously allows those students who have more experience to engage with classmates who have less experience in lab (typically biology majors guiding mathematics majors).

Each unit begins with a general biological introduction to entice biologists into the mathematics and coding while framing the basic biological questions for the mathematicians. The next and substantial portion of each unit is an introduction to the mathematical concepts and relevant coding. Each subsection is clearly labeled with the mathematical concept with the hope that it will be easy to go back to the ideas when needed. Each concept is applied to one or more biological problems, and exercises are scattered throughout. Some are thought problems, some best done by hand, and others require using R.

There are three cases per unit, each with a different biological scenario. Completing at least one will help solidify the critical concepts of the unit. After a list of goals for a given case, there is a short biological background or context. Next, the model is formulated, including the coding and data input with a series of step-by-step exercises. After we use the model to accomplish one goal, the case expands to either practice another mathematical skill or address a different question. Thus the cases can be done in full or in part or may be extended with data or questions more relevant to your situation. Once students have gotten started on one of the cases in class, we find it is feasible to assign completion of another case or two for homework.

The lab exercises introduce the fun and limitations of data collection. Why cannot everything be measured? Why are some data sets incomplete or messy? For this reason, each unit has a wet lab associated with it. Although these are found at the end of the init, they work well when done concurrently with the other activities in the unit since each might take up to three weeks to complete using a 2–3 hour lab period devoted to dry (prelab exercises, data analysis) and wet (collecting the data) activities. The first part of each lab develops the biological problems and includes a set of exercises to prepare students to analyze the anticipated data; these are handy even if you cannot do the lab itself due to time or access constraints. The lab exercises are likely to be somewhat familiar to thc biology students, but the modeling approach and integrating data with models will be new. There are exercises to guide students through the analysis both mathematically and conceptually. Moreover, since the outcome of applying a model to data is to generate new hypotheses, we suggest repeating the lab with new variables or designs suggested by the outcomes of the first iteration. Although we make suggestions, there is obviously extensive latitude for follow up activities. Indeed, there is extensive latitude for the lab activities themselves depending on available expertise and resources. Some other ways to achieve the lab goals are given in the "Technical Notes for Laboratory Activities" in Unit 5.

Each unit builds on the one before. That does not mean that it is impossible to simply select a case study from Unit 3 or 4 to use on its own, but, depending on student background, it may be necessary to provide more background on the modeling skills needed, often just the background for the unit itself.

Our goal is foster creativity and confidence in the ability of each student to address questions both experimentally and mathematically. We also hope students will learn to work together in interdisciplinary teams. To these ends, we suggest a final student designed project using their modeling skills, and if feasible, their lab skills, to address a question of their own design. Ideas for feasible projects are provided by the unit.

Tips for using this text – students

Our hope is that these lessons will indeed help you learn to create and use models of real biological problems. There are many things to learn simultaneously, including the math, coding in R, understanding the biology, and translating that to a plausible model. To develop the confidence to look at biological or other applied problems with these new insights, we have a few suggestions.

First, become familiar with the layout of the text. In Unit 1, we give the basics. The R Basics Section has all you need to get started with programming and R commands that appear repeatedly throughout the text. We suggest you do the exercises, and if you are an R wizard already, then perhaps offer to mentor a classmate who is new to programming. Learning R is best done with practice. We encourage asking each other for help; we all make mistakes when learning a new language, and trouble-shooting is part of what we are learning to do. And, you might note, the R Quick Help guide at the back of the book might be quite useful to you!

Second, with each unit the mathematics, modeling skills, and the biology become increasingly sophisticated. The first section of each unit develops these ideas through a series of stepwise exercises. We suggest you try them all even if they seem easy. Familiarizing yourself with this section of each unit will prepare you for the case studies or other applications of the mathematics you have just been learning.

Third, we are integrating biology, mathematics, coding, lab work as well as learning to identify questions and building a model to answer them. This can be daunting! The learning goals and specific goals for each exercise are intended to help you focus on each of the problems systematically.

Finally, these exercises, cases, and labs represent just a very few of the possible ways and problems to which mathematical modeling and the habits of mind associated with it might apply. We encourage you to develop your own questions and projects and use these skills in future classes and projects.

Online resources

The course companion website includes the following:

- data files:

List of data files available online

Section	Exercise	File name
1.2.7	Linear Regression Example	subantarctic.csv
1.2.7	Linear Regression: Data Transformation Example	ScalingBird.csv
1.2.7	Reading Data from Files, Exercise 1.16	SpecArea.csv
2.1.2	Calibrating Models using Linear Regression, Exercise 2.5	MothData.csv
2.2	Working through the Modeling Process, Exercise 2.6	Cranedata.csv
2.5.3	Garlic Mustard Case Study, Exercise 2.45	Fertility_Data.csv
2.5.3	Garlic Mustard Case Study, Exercise 2.46	Winter_survivorship.csv
3.4.4	Leaf Decomposition Case Study, Exercise 3.12	yellowbirch.csv
3.4.4	Leaf Decomposition Case Study, Exercise 3.15	redpine.csv
3.4.4	Leaf Decomposition Case Study, Exercise 3.15	redmaple.csv
3.6.2	Predator–Prey Case Study, Exercise 3.28	FRData.csv
3.6.3	Predator–Prey Case Study, Exercise 3.34	MysisData.csv

- all R listings throughout the book
- R Markdown templates for answering exercises
- suggested topics for small projects
- examples of successful final projects.

Acknowledgments

This project was inspired by Matt Richey, who encouraged the idea that we coteach a mathematical biology course with a wet lab. We have been supported by and taught with colleagues whose enthusiasm and guidance have played key roles in this project including Steve Freedberg, Steve McKelvey, Bruce Pell, and Hwayeon Ryu as well as our department chairs.

The project would not have been possible without help from numerous students who contributed to making it possible. Three students worked closely with us to draft the labs as part of a summer project supported by St. Olaf College and a TRiO McNair Scholarship: Nora Peterson, Megan Campbell, and Lansa Dawano were the first to bring student perspective and enthusiasm to this interdisciplinary project. Subsequent students contributed to developing some of the case studies, refining lab protocols, and developing supplemental materials. In particular, we thank Jesse Elder, Kaeli Jacobson, Ceci Sagona, Omar Sheta, Martha Sudermann, Jack Welsh, Julia Wolter, Megan Lapkoff, Joseph Rhoney, and Jonathan Deutsch.

We are also particularly grateful to Mike Swift for commenting on drafts of the manuscript and to Kevin Sanft for his guidance on teaching students R, and to their endless moral support in helping make this project a reality.

Students in our respective mathematical biology classes have taught us what is complicated and what reveals the greatest insights while building confidence and enthusiasm for the subject.

Finally, this project could not have happened without the community of quantitatively oriented biologists and biomathematicians who generated the projects we have cited and modified to build exercises and case studies based on real data.

UNIT

Preliminaries: models, R, and lab techniques

1

1.1 Bringing mathematics and biology together through modeling

At first glance, mathematics and biology might seem like unlikely partners. Mathematics uses formal logic to bring conclusive reasoning to abstract concepts. Biology, on the other hand, appears complex, unpredictable, full of exceptions and diverse levels of organization. However, biological systems are hierarchically organized to achieve a set of fundamental properties, such as the flux of materials and energy, that allow systems to reproduce, grow, and eventually die depending exquisitely on interactions among cells and tissues within an organism or individuals and species and their physical environments. These dynamic interactions are complex, requiring a myriad of processes at multiple levels of organization (e.g., molecular to the organism to the ecosystem scale). Mathematics allows us to develop abstractions of biological systems by using mathematical expressions to represent these processes. The act of abstraction, by translating these processes to a set of rules and studying how these processes affect the whole system, can help reveal new aspects of the relationship and thereby generate hypotheses and suggest experiments.

Models are generally defined as abstractions of reality, but the use of the word "model" varies greatly depending on the discipline. For example, a biologist may use a *conceptual* model, such as a flow diagram or concept map, to explain experimental observations, or an *animal* model to investigate human disease. A statistician may use a *statistical* model to identify patterns in data that a biologist then might use to formulate hypotheses. Here we will focus on mathematical models and how they can be used to gain insight into complex interactions in biological systems. In many ways, mathematical models are similar to flow diagrams intended to show the relationships (temporal, spatial) in a dynamic biological system. The addition of mathematical thinking and tools allows us to focus on the essential aspects of the system quantitatively, determine parameters that are otherwise inaccessible experimentally, understand the relative importance of various actors in a system, forecast future events, or predict reasonable consequences of interventions. Often, a model reveals that the concept map or flow diagram is incomplete leading to further exploration and novel understandings.

Mathematical modeling is itself an iterative process that uses mathematics to help scientists explore complex systems, make predictions, and generate causal explanations. Some models are built based on known relationships (e.g., exponential growth

Exploring Mathematical Modeling in Biology Through Case Studies and Experimental Activities
https://doi.org/10.1016/B978-0-12-819595-6.00007-4

or feedback inhibition), whereas others use mathematical relationships to best describe data sets and thus gain insight into the form of otherwise not fully understood relationships. Mathematical models are often too complex to find analytical solutions, and therefore we must use computers to simulate the models. These are often referred to as *computational* models.

Modeling is inherently collaborative. Modeling biological phenomena requires gathering empirical evidence and possibly the use of mathematical, statistical, and computational tools to test hypotheses, discover new insights, and suggest and refine experiments. The modeling process as outlined in Fig. 1.1 begins with an objective. When moving from the objective to building a mathematical model, it is often useful to draw a diagram that represents the relationships among variables and other factors that may influence the variables. These types of diagrams can be a common language between experimentalists and modelers. Like any concept map, these diagrams are dynamic, needing to be reorganized and refined as the different members of the team contribute new perspectives or as background research into the relevant literature changes the collective understanding of the problem.

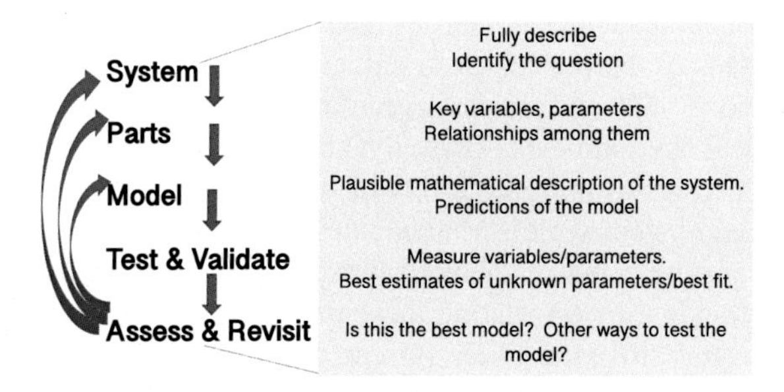

FIGURE 1.1

The iterative process of modeling.

The type of model used will depend on the questions being asked and the inherent properties of the system. Many different questions may be asked about a given system, and therefore it is important to clarify your particular goals before formulating a model. If the goal is to describe the relationship between variables and there is no need to explain why the variables interact the way they do, then a phenomenological or statistical model is sufficient. However, in this book, we focus on hypothesizing the interactions among variables and understanding what drives the observed behavior, and therefore we use mechanistic models.

Other questions we can ask to determine the type of model to use: Is the system static (at a steady state), or is it dynamic (changes with time)? Are the variables continuous, or do they only take on discrete values? Is it important to include how the variables change spatially? In this introductory book, we will introduce both discrete and continuous dynamic models that are deterministic, meaning that the output is fully determined by the parameters and initial conditions. Other texts might introduce stochastic models, which include inherent randomness.

Once a model is formulated, we can test the model predictions using actual data. A valid model can help us understand how a system works, predict a future state that is currently unknown, and determine what can be manipulated to produce a desirable output. Moreover, models often lead to questions that can be explored experimentally. Throughout this text, we will repeatedly work through the modeling process as depicted in Fig. 1.1, and our tools will become more sophisticated along the way.

For each of the cases and lab exercises, the modeling is motivated by a biological question and data, and the necessary mathematical and computational tools are introduced as needed. Throughout the text, the cases and lab exercises walk the reader through the modeling process, but the biological questions become increasingly more complex, and the mathematical and computational tools become more sophisticated. The text is by no means comprehensive. However, through this introduction to a logical quantitative approach to experimental design and analysis early in their academic careers, students will begin to develop an understanding of how the disciplines work together, how to communicate their ideas across disciplines, and the power and limitations of modeling and of experimentally derived data.

Using a computational environment is essential in mathematical modeling. Since we assume no prior programming experience, we begin by providing an introduction to the R programming language ("R Basics") focusing on the fundamentals of data entry and manipulation, graphing, linear regression, and iteration. Additional R skills are introduced as needed. Similarly, we provide the option of a very basic introduction to lab skills needed to succeed.

In Unit 2, we apply difference equations to a wide variety of biological problems with an emphasis on the modeling process: stating an objective, formulating a model, identifying assumptions, estimating model parameters from experimental data, interpreting model output, and using results to ask follow-up questions and propose new experiments. Linear regression is introduced and implemented to estimate parameters in several case studies (island biogeography, the pharmacokinetics of caffeine, and behavior of an invasive plant population) and to determine critical growth parameters from data on culture population sizes gathered in a wet lab.

Unit 3 introduces differential equations while continuing to expand our modeling toolkit. This unit focuses on formulating models, learning how to represent various biological relationships as mathematical expressions, and identifying the assumptions of models. We investigate the parameter space by creating sliders, finding best-fit parameters using nonlinear regression, and testing alternative models using the Akaike information criterion (AIC) for model selection. These methods are applied to a set of three case studies on the decomposition of organic matter, tumor growth, predator–prey interactions, and a wet lab on enzyme behavior.

In Unit 4, we introduce more complex systems of differential equations. Solutions are found numerically, and we implement optimization routines to estimate the best-fit values of parameters in the model given a set of observations. As part of assessing the applicability of each model, we learn to perform sensitivity analysis to study the effect of changes in parameters on the model output. These methods are applied to models of cancer immunotherapy strategies, infectious disease epidemics, quorum sensing in bacteria, and hormone-based homeostasis based on lab-accessible data.

Modeling requires a breadth of tools, including formulating mathematical equations to model living and inherently complex systems, applying statistical techniques to calibrate models, programming in R, interpreting results in the context of the biological question, and learning to evaluate the validity of a model. Each case study allows you to apply these tools and gain confidence in the process of modeling. A final few learning outcomes are to:

- Appreciate the value of organizing research by modeling.
- Realize that models help us access information or insights not possible or feasible from observation and experimentation.
- Enjoy the rich and occasionally frustrating process of interdisciplinary work.

Since we make no assumptions about prior skills in R or in lab, we introduce some of the basic skills that will be utilized throughout the text in the next two sections "R Basics" and the "Prelab Lab." We frequently reference the "R Basics" section for repeated core tasks such as uploading libraries, entering data, linear regression, for loops, and plotting. The "Prelab Lab" is designed to introduce basic lab skills and the practice of data analysis to students with little or no lab experience.

1.2 R basics

Computers are essential for visualizing data, solving discrete and continuous models, estimating parameters, and analyzing models. There are many programming languages that provide the capabilities to perform these operations. In this book, we will use R, an object-oriented scripting language that offers a rich set of built-in functions for statistical data analysis and modeling, excellent graphics functions for visualizing data and model output, and extensive documentation.

You can quickly install R for free on your personal computer by going to the appropriate link:

Windows: http://cran.r-project.org/bin/windows/base/

Mac: http://cran.r-project.org/bin/macosx/

For the Windows version, click on the download link at the top of the page. For the Mac version, click on one of the .pkg links to download.

We will be using RStudio, which provides a nice interface for using R. Go to http://www.rstudio.com. Click on **Download RStudio** and then choose **RStudio Desktop**. Select the installer for the correct platform. Now you are ready to go!

The best way to learn a language is to use it! It will be helpful to input the commands as you work through the tutorial. To obtain more information on any specific function in R, use the `?` help operator to access documentation for a function or package. For example, typing `?solve` will open documentation on the function `solve`.

Also, explore beyond what the case studies and labs dictate! Being willing to experiment will improve your ability to program and debug.

As you become comfortable with the RStudio layout, try to write up your results in an R Markdown document, which is easy to create in RStudio. R Markdown allows you to generate reports with embedded code and corresponding output, and it can export a file to html or pdf format. It is an excellent way to present your work as a reproducible record of your results. Refer to https://rmarkdown.rstudio.com for a tutorial. You can create your own R Markdown documents, or you can access templates for some of the background exercises and case studies from our book website.

1.2.1 RStudio layout

The first time you open RStudio, you will see three windows. The upper left window (see Fig. 1.2) is hidden by default but can be opened by clicking and choosing `R Script` on the tool bar at the top left of the page (this is also where you can choose R Markdown).

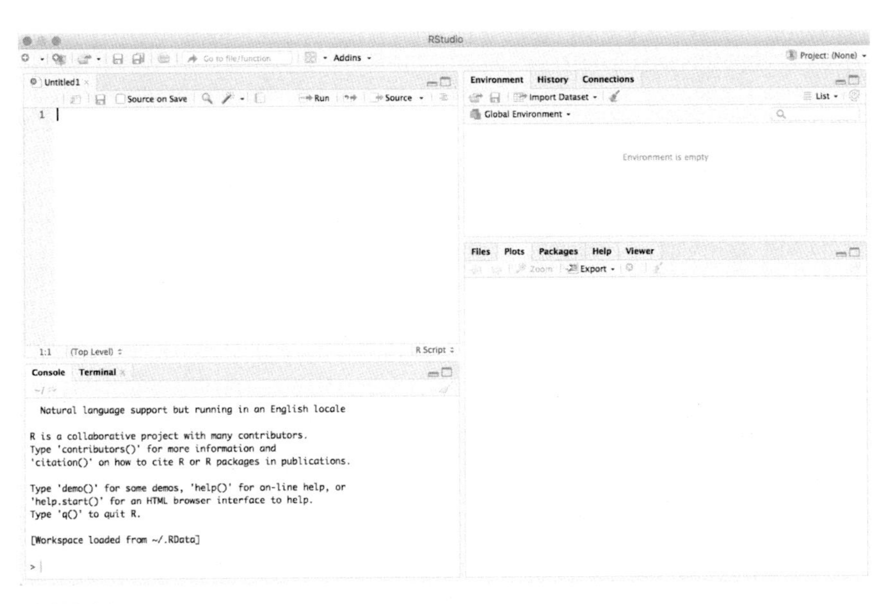

FIGURE 1.2

RStudio layout.

- The lower left is the *console window* (or command window). This is where simple commands can be entered following the prompt > and where output is printed.
- The lower right is the *files/plots/packages/help window*. Any plots that you generate will appear in this window. Moreover, by clicking on the various tabs at the

top of this window you can open files, use the help function, and install and load packages.

- The upper right is the *environment/history window*. Here you can view objects in your working space (under the environment tab) or view your command history (under the history tab).
- The upper left is the *script window*.

Libraries

R uses packages, which are collections of functions. Many packages are installed in the standard installation of R, but throughout this book, you will need additional packages. For example, in Unit 3, we will use the package `manipulate`, which allows us to create plots with sliders. To install this package, type the command `install.packages("manipulate")` in the console. You will only need to do this once on your computer, but once a package is installed, it must be loaded in each new R session using the command `library(manipulate)`.

Scripts

Simple calculations can be entered in the console window, but we often want to organize a collection of commands that can be edited and saved. This can be done using a script file. If you have not already opened a script file, go to ⊕ in the script window and choose `R Script`. Once you have typed commands in this space, there are various ways to run the code. You can use the `Run` toolbar button or Ctrl+Enter key to run the line where the cursor resides, or you can run a section of the code by highlighting part of it and then running it. You can also use the `Source` toolbar button to run the entire script.

R will ignore any commands following the symbol #. This feature allows us to include comments in a script, and it is extremely important to comment your code well! Including comments will help you understand your code if you come back to it at a later time, and it will be easier for someone else to read and understand your code.

As you work through the examples presented throughout this tutorial, try entering the commands in either a script file or an R Markdown document.

Exercise 1.1

What does the built-in function `log` do? Try reading the help page by entering ?log in the console window and/or do a web search (use key words such as "R log function") to learn about the function. In R, use this command to calculate $\ln 4$ and $\log_{10} 4$.

1.2.2 Simple calculations

At the command prompt > in the console window, type `5^2-2` and hit Enter. You will see

```
> 5^2-2
[1] 23
```

R follows basic order of operations, but parentheses can be used to force a different order. R also has an extensive library of built-in functions. The exponential function is `exp(x)`, the square root is `sqrt(x)`, and the natural logarithm is `log(x)`. Refer to the R cheat sheet in the appendix for a list of commonly used functions.

You can also assign numbers to variables. For example, when we type `x=2`, the variable `x` has been created and assigned the value 2. R now remembers the value of `x`, and we can do calculations with `x`:

```
> x=2; 5*x+4
[1] 14
```

Note that a semicolon allows two commands to be typed on a single line. Once a variable is defined, you will see it appear in the environment panel (upper right panel). To remove all variables from R's memory, type `rm(list=ls())`. You can also click on the broom icon in the environment window to clear all objects.

Exercise 1.2

(a) Compute `-2^4` in R and compare to `(-2)^4`.
(b) Compute $(\sin(\pi/5))^2$ in R.
(c) Consider the following input in R:

```
a=2; a=a+5
```

What number is now assigned to `a`?
(d) Consider the following input in R:

```
x=5; y=6; x=y
```

What are the values of `x` and `y` after running this code?

1.2.3 Data structures

R supports basic data structures, which include vectors, matrices, and data frames. *Vectors* are ordered collections of elements, in which each of the elements must be of the same type (e.g., numeric, character). A *matrix* is a rectangular array with some numbers of rows and columns where each element must be of the same type. Similar to a matrix, a *data frame* is also a two-dimensional array, but each column can be of a different type. Here we provide a brief overview of how to construct these data types and how to address elements of them.

Vectors

Vectors can be explicitly constructed using the concatenation function `c()`:

```
> c(1,7,3,4)
[1] 1 7 3 4
```

Another useful way to construct a vector is to use the sequence function:

```
> seq(0,1,by=0.2); seq(0,1,length.out=6)
[1] 0.0 0.2 0.4 0.6 0.8 1.0
[1] 0.0 0.2 0.4 0.6 0.8 1.0
```

Note that the first input of `seq` is the initial element and the second input is the last element. We can use `by` to specify the increments, or we can use `length.out` to generate a vector of a set number of evenly spaced values. There are also shortcuts for creating vectors that increment by 1:

```
> 1:5; seq(1,5)
[1] 1 2 3 4 5
[1] 1 2 3 4 5
```

The `rep` function creates a vector with replicates of a value or values. The basic syntax is `rep(x, times)`, where `x` can be a value or a vector (of any type). For example,

```
> rep(0,4)
[1] 0 0 0 0
```

Elements in vectors can be addressed by square bracket `[i]` indexing, where indexing starts at 1. For example, suppose we define a vector `v` and we would like to access the fourth element:

```
> v=c(10,1,8,9,22,3); v[4]
[1] 9
```

We can also access part of a vector. For, example, to access the third through fifth elements of `v`,

```
> v[3:5]
[1] 8 9 22
```

To determine how many elements are in the vector `v`, we use the `length` command:

```
> n=length(v)
[1] 6
```

Now that the length of the vector is defined, `v[2:n]` accesses all elements of `v` except for the first element, but a more efficient method to delete the first element of the vector is

```
> v[-1]
[1] 1  8  9 22  3
```

Note that the vector `v` still contains all original elements. If we assign `w=v[-1]`, then this new vector now contains all elements of `v` except for the first element. Experiment with deleting other elements of this vector (e.g., `v[-n], v[-(1:3)],v[-c(2,4)]`).

Exercise 1.3

Create a vector **u** that contains all the integers from 10 to 40, inclusive. Then write a one-line command to access a vector consisting of the tenth, first, and third elements of u in that order.

Exercise 1.4

Create a vector **v** that contains all even years from 1910 to 2000, inclusive. Then, using **v**, create a new vector that includes the tenth element through the last element of **v**.

Matrices

A matrix is a two-dimensional array of numbers. A matrix can be entered by creating a vector of the matrix entries and then specifying the desired number of rows and columns using the matrix function. For example,

```
> X=matrix(c(1,2,3,4,5,6),3,2)
```

produces the 3 by 2 matrix

```
> X
     [,1] [,2]
[1,]    1    4
[2,]    2    5
[3,]    3    6
```

Sometimes we may want to combine two vectors of the same length. The function `cbind` binds two vectors to form a matrix:

```
> M=cbind(1:3,4:6); M
     [,1] [,2]
[1,]    1    4
[2,]    2    5
[3,]    3    6
```

Suppose we want to access an element of `M`. This is similar to addressing an element of a vector, but now we need to specify the row and column. For example, to access the element in the 3rd row and 2nd column,

```
> M[3,2]
[1] 6
```

We can also access an entire row or column. To do this, leave either the row limit or column limit blank. For example, suppose we want to access the first column and assign it to `u`. This means we want all rows and only the first column,

```
> u=M[,1]; u
[1] 1 2 3
```

Now to set the third row equal to r,

```
> r=M[3,]; r
[1] 3 6
```

Exercise 1.5

The function rbind works similarly to cbind, but it combines by rows instead of columns. Use rbind to add a fourth row with elements (9, 13) to the matrix M defined in this section.

Exercise 1.6

Construct a 5×5 matrix B with elements that are the numbers 1 through 25 (reading from left to right, top to bottom. (Type ?matrix to learn how to fill in a matrix by rows.) Access its second column.

Data frames

A data frame is similar to a matrix, but each column can be of a different type (numeric, character, etc.). This can be very useful, but throughout this book, we will typically use data frames that include only numeric types. Another nice feature of a data frame is that the columns have names, and we can access a column by its name. For example, we can construct a data frame A,

```
> A = data.frame(time = c(0,3,4,6), N = c(11,15,16,18)); A
  time  N
1    0 11
2    3 15
3    4 16
4    6 18
```

which contains two columns named time and N. We can select the first column by using A[,1], or we can select the first column using its name:

```
> A$time
[1] 0 3 4 6
```

Throughout this book, we will often import data, and these data will be read in as data frames. Data in the form of a data frame can then be used and manipulated in various ways as shown in the next exercise.

Exercise 1.7

Create three vectors of your choice, and then create a data frame from the three vectors that you defined. Find the mean of each column of the data frame by calling up the data frame name and column name. (Type ?mean for help on using this function.)

1.2.4 Basic plotting

R is used by many data analysts because of its graphic capabilities. In this book, we will only need basic graphs, and therefore we will use the `plot()` function. However, interested students can investigate other plotting packages such as `ggplot2`.

To get a feel for plotting in R, consider plotting the function $f(x) = \sin x$ for x between 0 and 2π. First, create a vector of values for x, and then plot the function:

```
x=seq(0,2*pi,length.out=20)
plot(x,sin(x))
```

If `x` is a vector, then `sin(x)` will evaluate the sine function at each element of the vector (try typing `sin(x)` in the console and view the output). In the plotting command the independent variable comes first, followed by the dependent variable. Try putting these commands into RStudio. Run the code, then change the length of the vector `x` to 100. What happens?

Note that open circles are the default option for the plot command. Try `plot(x,sin(x),pch=2)`. What is different about your plot? Try using a different number in the pch option. If we want a line graph instead of a scatter plot, then we can include the `type` option in the plot function, `plot(x,sin(x),type="l")`. Another curve can be added to the existing plot using the `points` or `lines` command. Run `points(x,cos(x))`, and then change this to `lines(x,cos(x))` and observe the difference.

Exercise 1.8

In R, plot $y_1(t) = \frac{t^2}{1+t^2}$ and $y_2(t) = \frac{t}{1+t}$ for $t = 0, 0.1, 0.2, ..., 4.9, 5$ (t from 0 to 5 in increments of 0.1). Plot both $y_1(t)$ and $y_2(t)$ on the same graph as line graphs.

In this text, we will often want to plot data. Consider the honey bee life history data in Table 1.1 on the populations of eggs, larvae, pupae, and adults [24]. We will first plot the number of eggs over time and then overlay the other data. The first step is input the data into R as vectors:

```
t = seq(12,180,by=12)
eggs = c(2586,2401,3341,4595,4977,4864,4162,2220,3916,2612,
            1373,23,0,0,0)
```

Plot eggs versus time,

```
plot(t, eggs, pch=15)
```

where plot character 15 (filled square) is used. We can add a title and specify a label for each axis:

```
plot(t, eggs, pch=15, xlab="Days after hiving",
    ylab="Population (individuals)",
    main="Honey Bees Population", ylim=c(0,46000))
```

Table 1.1 Numbers of eggs, larvae, pupae, and adult bees. Data from [10].

Days after hiving	Eggs	Larvae	Pupae	Adults
12	2586	3539	0	5362
24	2401	3657	8919	3419
36	3341	5089	8281	9629
48	4595	6998	12293	17259
60	4977	7579	13137	23412
72	4864	7408	17161	29185
84	4162	6339	16772	36829
96	2220	3380	14353	43186
108	3916	5964	7654	45003
120	2612	3978	13503	37778
132	1373	2091	9008	35332
144	23	34	4734	30495
156	0	0	78	19674
168	0	0	0	12704
180	0	0	0	6438

Try changing the values in the `ylim` option. What happens? You should see that the `ylim` option defines the lower and upper limits for the vertical axis. R sets the vertical scale in the first plot command, and therefore the upper limit was chosen so that the adult data will appear when we add it to this existing plot. We can add the larvae, pupae, and adult data by defining a vector for each and then using `points`:

```
larvae = c(3539,3657,5089,6998,7579,7408,6339,3380,5964,3978,
              2091,34,0,0,0)
pupae = c(0,8919,8281,12293,13137,17161,16772,14353,7654,13503,
              9008,4734,78,0,0)
adults = c(5362,3419,9629,17259,23412,29185,36829,43186,45003,
              37778,35332,30495,19674,12704,6438)

points(t, larvae, pch=16, col="blue")
points(t, pupae, pch=17, col="orange")
points(t, adults, pch=18, col="green4")
```

Different plot characters and colors are included in the plot command to distinguish the various data sets. To make the graph easier to read, a legend should also be included:

```
legend("topleft", c("eggs","larvae","pupae","adults"),
         pch=c(15,16,17,18), col=c("black","blue","orange","green4"))
```

The location of the legend, the labels, the plot characters, and the colors have all been defined in the legend command. The result is shown in Fig. 1.3. To save a graph, go to Export and choose Save as Image or Save as PDF. Type `?plot` to learn more about how to customize plots.

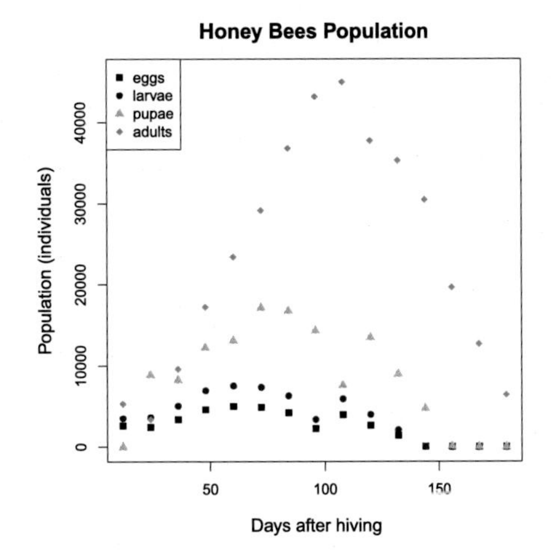

FIGURE 1.3

Number of brood and adults. Data are listed in Table 1.1.

Exercise 1.9

Plot the number of brood (the sum of the number of eggs, larvae, and pupae) versus time and overlay the number of adults over time. Include a legend and make sure the plot character and color correspond to the correct set of data in the plot.

Exercise 1.10

Data describing life expectancy of human males and females at birth are given in Table 1.2 (The World Bank). Use these data in this exercise.

Table 1.2 Life expectancy at birth for males and females in the Unites States. *Source: The World Bank.*

Year	Females	Males
1960	73.1	66.6
1965	73.8	66.8
1970	74.7	67.1
1975	76.6	68.8
1980	77.4	70
1985	78.2	71.1
1990	78.8	71.8
1995	78.9	72.5
2000	79.3	74.1
2005	80.1	75
2010	81	76.2
2015	81.2	76.3

(a) Make a vector for each column of data in Table 1.2 (or import Life_Expectancy.csv), and plot female life expectancy versus year. Replicate the plot shown in Fig. 1.4A.
(b) Now plot the data for females and males on the same graph. Replicate Fig. 1.4B.

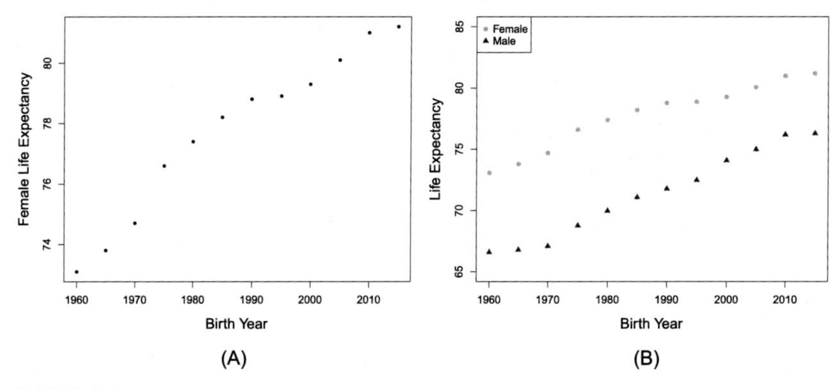

FIGURE 1.4

(A) Life expectancy plot for Exercise 1.10(a). (B) Life expectancy plot for Exercise 1.10(b).

1.2.5 Reading data from files

Throughout this book, we typically import data rather than typing data into RStudio. If your data is in a .csv file, then you can use the command `read.csv` to import the data (or `read.xlsx` for an Excel spreadsheet). The advantage of importing data using these commands is that the step to import data is documented in your script, and this method can also be easier when it is necessary to import multiple or large files. However, RStudio has an alternative method that works as well. The data import features can be accessed from the environment panel, as shown in Exercise 1.11. The next exercise leads you through importing a data set (see Fig. 1.5).

Exercise 1.11

In this exercise, we will download the data set SpecArea.csv. To do this, go to "Import Dataset", choose "From Text (base)..." and then find the file SpecArea.csv and choose Open. A window will appear that allows us to choose among some options. Note that it will automatically fill in the name as the file name, but you are free to change this. Also, note that it asks if a Heading is included. The first row of the csv file does in fact include column names, and therefore you will leave Yes marked. Once you import the data, type `head(SpecArea)` to see the first six rows of data. The data are now in the form of a data frame. Recall that a data frame allows us to call up each column by the column name, that is, `SpecArea$area` and `SpecArea$species`. Plot the number of species versus area. On a separate plot, plot the natural log of the number of species versus the natural log of the area. What do you observe?

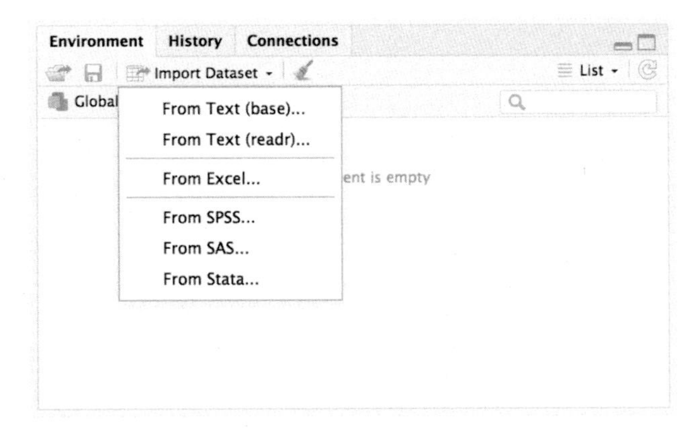

FIGURE 1.5

Import Dataset tab found in the Environment Window in RStudio. Choose "From Text (base)" to import a csv file.

1.2.6 Iteration

Loops can be used to automate the same operation repeatedly. Here we focus on `for loops`, which perform some operation for a specified number of times. We will start with a simple example. Suppose we want to print the integers 1 through 5. A `for loop` can be used to do this:

```
for (i in 1:5) {
  print(i)
}
```

In this example the command within the curly brackets is run five times. When we first enter the loop, `i=1`, and we print this value. The second time through the for loop, `i=2`, and we print this value, and so on.

In this book, there are many examples where we are performing a repeated operation, and we want to store the result in a vector. Again, consider a simple example where we would like to create the vector $\mathbf{y} = (1, 2, 3, 4, 5)$. There are simpler ways to do this, but here we will use a *for loop*. We first initialize a zero vector that will later be used to store assigned elements. To create a zero vector $\mathbf{y}$ with five elements, recall that the `rep` function can be used to do this:

```
y=rep(0,5)
[1] 0 0 0 0 0
```

In the *for loop*, the counter `i` will run from 1 to 5 to fill in the elements of `y`:

```
for (i in 1:5) {
  y[i]=i
}
y
```

The output is

```
[1] 1 2 3 4 5
```

When we first enter the loop, `i=1`, the first element of **y** is assigned the value `i`, that is, `y[1]=1`. The second time through the loop, `i=2`, and we set `y[2]=2`, and so on.

Suppose a population in year 1 has 50 individuals and the population grows 10% each year. Let `p[i]` represent the population in year `i`. Then `p[1]=50`, and we can calculate the population in the next year using `p[2]=1.1*p[1]`. The following year, `p[3]=1.1*p[2]`, and so on. There is a pattern, and we see that the population in year `i+1` is written in terms of `i`, that is, `p[i+1]=1.1*p[i]`. We can use this repeated calculation in a `for loop` to compute the population from year 1 to year 10:

```
p=rep(0,10)
p[1]=50
for (i in 1:(length(p)-1)) {
  p[i+1]=1.1*p[i]
}
p
```

We could have typed `i in 1:9` for the counter, but it is helpful to get into the habit of writing things in terms of things already assigned. This way, if we decide to compute the population for 20 years instead of 10, then we will only need to edit one line.

Exercise 1.12

In the previous example, based on the fact that the vector `p` is initially defined as `p=rep(0,10)`, explain why the counter for `i` goes to `(length(p)-1)` instead of `length(p)`.

Exercise 1.13

Consider the previous example again, where the initial population is 50, and the population grows 10% each year. Change the code so that the counter is `(i in 1:length(p)-1)` and run the entire `for loop` (note the differences in parentheses!). What happened? Can you explain the outcome? (If you are not sure how to explain this, then try just running `1:length(p)-1` and `1:(length(p)-1)` in the console. What are the differences? Why would one of these result in an error for the operation you are performing in the `for loop`?)

Exercise 1.14

Suppose a population in year 1 has 10 individuals and the population decays by 5% each year. Write a `for loop` to compute the population from year 1 to year 20.

We can also have a `for loop` inside another `for loop` (nested for loops). For example, consider the previous example in which a population is growing by 10% each year. Suppose we want to calculate the population over 10 years but for initial populations 5, 10, 15, and 20. We can create a vector of initial conditions and insert another `for loop` around our previous `for loop`:

```
m=10   # number of iterations for time
init=c(5,10,15,20)   # initial values of p
n=length(init)   # number of initial values
p=matrix(0,m,n)   # zero matrix with m rows and n columns
p[1,] = init   # store initial populations in first row of p

for (j in 1:n) {   # iterate for each initial poplation
  for (i in 1:(m-1)) {   # fill in column j with population over time
    p[i+1,j]=1.1*p[i,j]
  }
}
p
```

Previously, wc created a vector p that stored the population values for each of the ten years. Here we want to do the same calculation but for four different initial population values. Therefore we can set up a 10 × 4 matrix where each column will store the population over 10 years for a particular initial condition. The outer loop cycles through the four initial conditions, and the inner loop computes the population over ten years. Copy this code and run it in R. Take time to understand what each line is doing! It can be helpful to write the first few iterations with pencil and paper, that is, set j=1 and cycle through the inner loop for i=1, i=2,... and write the output. Loops will be used throughout this book, and therefore it is essential to understand them now!

1.2.7 Fitting a linear regression model

Regression analysis is a statistical approach to modeling the relationship between one or more input variables and one or more output variables. The simplest form of regression, called simple linear regression, involves finding the linear relationship between one input variable x and one output variable y. Linear regression will be used in Unit 2, but we will give a brief background here. We also introduce nonlinear regression in Unit 3.

We have all encountered the equation of a line, $y = mx + b$, where m is the slope, and b is the intercept where the line intersects the vertical axis. In the real world the relationship between two variables will never be perfectly linear, but the relationship might be close to linear. For example, consider the data shown in Fig. 1.6A [12]. In this study, data on reproductive biomass and the total number of seeds were collected from *Colobanthus quitensis*, a flowering plant, in South Georgia (sub-Antarctic). Suppose we want to model this relationship using a linear equation given by

$$y = b_0 + b_1 x, \tag{1.1}$$

where in this example x represents the reproductive biomass, and y represents the total number of seeds. The parameter b_0 is the estimated intercept, and b_1 is the estimated slope that best fits the data.

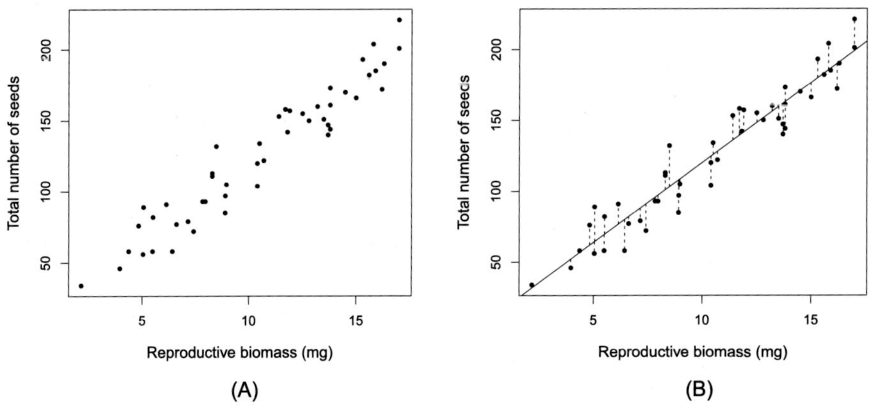

FIGURE 1.6

(A) Data on reproductive biomass and number of seeds for *Colobanthus quitensis* collected at various sites in South Georgia (subset of data extracted from Fig. 1 in [12]). (B) Line of best fit (solid line) and residuals (dashed lines).

The data can be found in subantarctic.csv and then imported into RStudio. The `lm()` function will fit a linear model to the data:

```
> fit.line=lm(y~x)
```

where `x` and `y` correspond to the input and output variables, respectively. Note that when we use the command `lm`, the output variable (`y` in this case) comes first, and the input variable (`x`) comes second, that is, `lm(y~x)`. The results are saved to an object we have named `fit.line`. If we type `fit.line` and run this, then we see the following output:

```
> fit.line

Call:
lm(formula = y ~ x)

Coefficients:
(Intercept)            x
      7.558       11.257
```

The `lm` command established the relationship between the predictor x and response y, that is, it determined b_0 and b_1 given in Eq. (1.1) that best fit the data. These values are found in the `Coefficients` part of the output. The estimated intercept b_0 is 7.558, and the estimated slope b_1 is 11.257. In other words, it found the line of best fit to be approximately $y = 7.56 + 11.3x$. We can plot the data and overlay the line of best fit using the function `abline()`:

```
plot(x,y)
abline(fit.line)
```

The output is shown in Fig. 1.6B, where the distance between each data point and the line are displayed. This difference is called the *residual* and is defined as

$$e_i = y_i - (b_0 + b_1 x_i) \tag{1.2}$$

The residuals can be read by typing `resid(fit.line)`. These residuals represent the differences between the data and model, and the line of best fit is determined by finding the values of b_0 and b_1 that minimize the sum of the squared residuals, which will be discussed further in Unit 3.

Now that a linear model is found, we may want to compute summary statistics to see whether the two measurement variables are associated with each other and estimate the strength of the relationship between the two variables. For detailed summary statistics of the linear fit, type `summary(fit)`. The summary gives the p-value for the estimated slope, which is a test to see if the estimated slope is significantly different from zero. If the p-value is greater than 0.05, then we should question whether there is a linear relationship between the input and output variables. The output also gives the R-Squared statistic R^2, which is the proportion of the variance in the dependent variable explained by the independent variable. This value can vary from 0 to 1, and it expresses the strength of the linear relationship between the input and output variables. An R^2 value close to 1 means there is a strong linear relationship between the variables.

When using linear regression, it should be noted that there are four assumptions about the distributions of residuals. First, the residual associated with one output value has no effect on the residuals associated with other output values, that is, the residuals are independent. Knowledge of the study design or data collection must be known to verify that this assumption is met. Second, the probability distribution of the residuals is normal for a particular input value. A normal probability plot can be used to check this (not shown here), but if we do not have multiple experiments for a given input value, then this can be hard to verify. Third, the mean of the residuals for a given input value is zero, and fourth, the variation of observations around the regression line is constant. The residual plot, a scatter plot of the residuals versus the fitted output variable, is a helpful tool in determining if these last two assumptions are satisfied. The residual plot for the data and linear model in Fig. 1.6B is shown in Fig. 1.7. To produce this plot, we first need to obtain the fitted y values, which means that we will plug the x values into our linear model. To access the estimated slope and intercept, type `fit.line$coefficients`. You should see that the estimated intercept is the first element, and the estimated slope is the second element. Therefore we can compute the fitted values

```
y.fit = fit.line$coefficients[[1]] + fit.line$coefficients[[2]]*x
```

and then plot the residuals versus these fitted values:

```
plot(y.fit,resid(fit.line), xlab="fitted values",ylab="residuals")
```

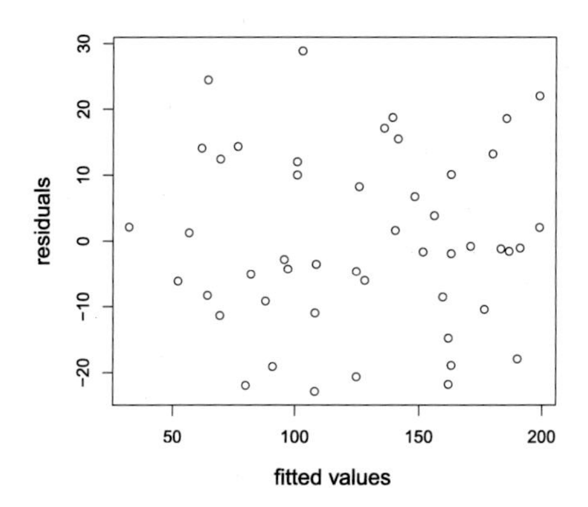

FIGURE 1.7

A plot of the residuals (data and linear model shown in Fig. 1.6B) versus the fitted values for numbers of seeds.

If the variables follow a linear relationship, then the residuals should look like they bounce around a mean of zero. If the variation in residuals is constant, then the residual plot should look like a cloud of points, and we should not observe any pattern. Both of these conditions seem to be met based on the residual plot in Fig. 1.7. Any trend in the residual plot might suggest a nonlinear relationship between the two variables. However, note that we will often use small data sets throughout this book, which means it can be difficult to show that these assumptions are met!

Exercise 1.15

Consider the life expectancy data for females in Table 1.2 and the corresponding plot in Fig. 1.4A. There are twelve data points that we will denote by (t_i, y_i) where $i = 0, ..., 11$. We might suggest that during this time period, there is approximately a linear relationship between birth year and life expectancy. Use linear regression to find the best-fit line through the data. Plot the data and overlay the linear model. In a separate plot, graph the residuals versus the fitted value. Discuss your conclusions.

If there is no strong linear relationship between two variables, sometimes a linear equation is appropriate if the data are transformed (e.g., taking the log of one of the variables). For example, consider the data (found in ScalingBird.csv) on the mass (g) and field metabolic rate FMR (kJ/day) of birds [45] shown in Fig. 1.8A. By visual inspection it seems there is no strong linear relationship between mass and FMR. We could attempt to fit a linear model and use the residual plot to confirm this (try this!). We might try transforming one or both of the variables in an attempt to find a linear relationship. However, in this example, we want to describe how a characteristic, FMR in this case, changes with size.. The study of these scaling relationships is referred to as *allometry*. Researchers have looked at many different scaling relationships and have noticed that they often look linear when plotted on a log-log scale, and therefore the relationship between the original variables can be represented by a

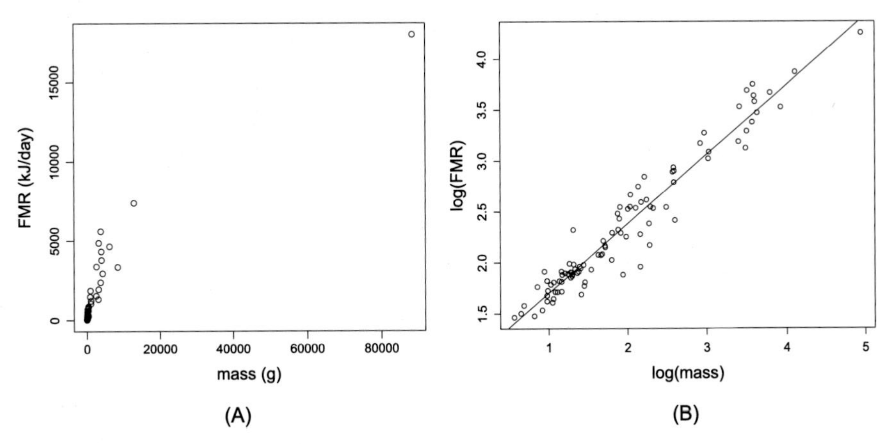

FIGURE 1.8

(A) Field metabolic rate in kilojoules per day versus mass of birds. Data from [45]. (B) Line of best fit (solid line) after log transformations of both variables.

power law given by

$$f = ks^a, \tag{1.3}$$

where s is mass, and f is FMR in our example. The parameters k and a are constants, and a is called the allometric coefficient. If the data closely follow this relationship, why do the data on a log-log plot follow a linear pattern? If we take the log of both sides of Eq. (1.3), then we obtain

$$\begin{aligned} \log f &= \log(ks^a) \\ &= \log k + \log(s^a) \\ &= \log k + a \log s. \end{aligned} \tag{1.4}$$

This is the equation of a line! It looks different than the general form of a line, $y = mx + b$, but if we think of $\log f$ as our output variable ("y") and $\log s$ as our input variable ("x"), then a is the slope, and $\log k$ is the y-intercept. The result of plotting the log of the metabolic rate versus the log of the mass is shown in Fig. 1.8B. By visual inspection there seems to be a linear relationship, which means the power law equation $f = ks^a$ might be a good model for metabolic rate as a function of mass.

To produce Fig. 1.8B, import the data file ScalingBirds.csv, then take the log of each variable and plot them:

```
x = log(ScalingBirds$Mass,10)
y = log(ScalingBirds$FMR,10)
plot(x,y,xlab="log(mass)",ylab="log(FMR)")
```

To fit a linear model, perform linear regression using the transformed variables and plot the result:

```
fit.line = lm(y~x)
abline(fit.line)
```

If we want to see only the estimated slope and intercept, type `fit.line$coefficients`. Assign these components to variable names:

```
intercept = fit.line$coefficients[[1]]
slope = fit.line$coefficients[[2]]
```

Using Eq. (1.4), we know that the slope is equal to a and the intercept is equal to $\log k$. Therefore

```
a = slope
k=10^intercept
```

that is, $a \approx 0.681$ and $k \approx 10.52$. Therefore Eq. (1.3) can now be written as $f = 10.52s^{0.681}$.

It should be noted that data transformations should be used with caution! For example, if we take the reciprocal of a variable, then small errors can become large errors. This will be discussed further in the enzyme lab in Unit 3 when we compare the results of regression analysis using various transformations. For a more thorough discussion on regression analysis assumptions and data transformations, see [42].

Exercise 1.16

Understanding patterns of spatial diversity has important implications for conservation of biodiversity. It is well known that the number of species within a taxonomic group increases with island area. Terborgh (1973) used birds to investigate the species-area relationship in the West Indies (see Table 1.3). Here we will use this data set to understand the relationship between area and number of species. The species-area relationship can be approximated by a power function of the form

$$S = cA^z,$$

where the constants c and z are specific to a data set. This power function can be represented in linear form:

$$\log S = \log c + z \log A, \tag{1.5}$$

where $\log A$ is the independent variable, $\log S$ is the dependent variable, z is the slope, and $\log c$ is the vertical intercept. (Note that we could have also used the natural logarithm to transform the equation to a linear equation. However, most biologists use $\log_{10}$ as it is easier to interpret the raw data by powers of 10.)

(a) Import SpecArea.csv, which contains the data on area and bird species numbers listed in Table 1.3. Plot the log of the number of species versus the log of the area.
(b) Perform linear regression on the linear form of the species-area relationship (Eq. (1.5)) and plot the regression line on the same plot as the data.
(c) For this particular data set, what are the estimates of c and z?

Table 1.3 Area and bird species numbers in the West Indies [68].

Island	Area (mi^2)	No. of species
Cuba	44,164	79
Hispanioloa	29,979	79
Jamaica	4,411	68
Puerto Rico	3,423	54
Guadeloupe	583	34
Martinique	385	41
Dominica	305	40
St. Lucia	233	44
Barbados	160	16
St. Vincent	133	38
Grenada	120	22
Antigua	108	23
St. Croix	80	29
Grand Cayman	71	29
St. Kitts	68	21
Barbuda	62	22
Montserrat	38	24
St. Martin	33	17
St. Thomas	32	24

Make sure you are comfortable with all of the commands introduced in this tutorial before working further through this text. These commands will be used repeatedly throughout the text, and you will also be introduced to new commands as needed. Refer to the `R Quick Help` sheet in the back of this book as a quick reference for commonly used commands.

1.3 Prelab lab: practicing the fundamentals

Goals: The primary goal of this lab is to become familiar with two tools we will use, micropipettes for accurately measuring small volumes and spectrophotometers for determining concentrations based on light interactions with a solution. Along the way, we will practice safe liquid handling techniques. We will record data, plot data, and analyze the data using linear regression. This lab covers basic lab and math skills that will be used repeatedly throughout this book.

Introduction

During our experiments, we will frequently need to transfer "an aliquot" or small volume of liquid from a stock solution to our experimental mixture or to a tube for additional measurements or storage. To do this, we commonly use micropipettes with disposable tips. The volumes range from 1–2 μL to 1 mL with different pipettes and tips designed for smaller and larger volumes, respectively. They are typically adjustable allowing measurements of any volume within range (e.g., 45 μL). Using them is easy with just a little practice, and if you pay attention, then they are extremely accurate. We will start by getting familiar with the pipettes and tips. Then we will use them to prepare a serial dilution of a blue solution. Dilutions using known volumes allow preparing a more dilute solution from a concentrated one; for example, you may have a stock solution of 1 mg/mL albumin, but you need 200 μL of 125 μg/mL.

Exercise 1.17

(a) Describe how you will prepare 200 μL of 125 μg/mL from a 1 mg/mL stock solution and distilled water. Show your calculations with units. Explicitly list the volumes of stock solution and distilled water that must be combined. Explain how you could do this with the micropipettes we have.

(b) How would you prepare 1 mL of 5 μg/mL albumin? Show your work and indicate the volumes needed. Is this dilution feasible to prepare with our pipettes? Can you think of a way to actually do this?

As we see in the second problem, sometimes it is necessary to make a huge dilution step, but the pipettes are good only in a limited range. Sometimes we need to make a series of concentrations, each 4-fold or 10-fold smaller than the other. In both of these cases, serial dilutions or a sequence of dilution steps is more accurate than pushing the limits of our pipettes. We will try this!

Another common lab problem is determining the concentration or density of a solution. Fortunately, if a compound absorbs light in the ultraviolet or visible range, then we have a handy instrument that will measure the extent of light absorbance, the spectrophotometer. For example, anything with color has absorbed some light in the visible spectrum. The amount of light absorbed is directly proportional to the concentration of the absorbing chemical in solution, that is, if the concentration is doubled, then the absorbance doubles. The output is reported as absorbance at a particular wavelength and is accurate (depending on the measurement) from 0–1 or 1.5.

We can use the same instrumentation to measure how turbid a solution is. Turbid solutions like whole milk or bacteria in suspension scatter light in multiple directions deflecting light away from the photomultiplier tube. The spectrophotometer, which is designed to report the amount of incoming light that actually reaches its photomultiplier tube, will register more scattering due to, for example, more bacteria, as greater absorbance; we often distinguish scatted light from absorbance by calling it optical density (OD). Typically, absorbance or OD measurements are done at a defined wavelength and in special containers with flat sides called cuvettes to minimize the light scattered by the container and to ensure that the length that the light travels through the solution is always the same (typically 1 cm). Thus to use a spectrophotometer, it is important to decide on and set the optimal wavelength and to adjust the instrument to "zero" with a solution that is identical to the one being tested, just minus the dye or bacteria. We will practice doing this with our serial dilutions.

Spectrophotometers come in all shapes and sizes and are extremely accurate. Many labs are replacing their cuvette-style spectrophotometers with an absorbance plate reader that uses similar optics to measure absorbance, but instead of stand-alone cuvettes, samples are put in small wells arranged in a plate. The number of wells can vary, but 96 per plate is fairly typical. If this is the way your spectrophotometric measurements will be made, then everything will be the same except that your final dilutions will be transferred to individual wells in a plate instead of a cuvette.

Before we get started, identify all the materials in the list in the following section. Note that today all the solutions are very safe, water and water with food coloring so that the biggest concern is to avoid getting food coloring on your clothes. However, we should practice good lab rules about liquid and tip disposal (note specified containers) and report any spills to your TA or instructor.

Materials

- Micropipettes such as P1000, P200, P20 and tips.
- 5-mL tube of water with a nontoxic coloring agent (blue food coloring) with the concentration adjusted to an absorbance of about 1 at 410 nm.
- Parafilm squares.
- DI water.
- Test tubes (e.g., 10 × 70 or 13 × 100), 6–7 per group.
- Test tube racks.
- Markers/tape for labeling.
- Container for liquid waste and a waste container for tips.
- Vortex or tube caps/parafilm for mixing tubes by inversion.
- Spectrophotometer or plate reader for absorbance measurements.
- Semimicrocuvettes (1 mL), 6–7 per group or 48- or 96-well plates.

Using the micropipettes

Micropipettes with matching disposable tips are used to measure small volumes of liquid. The P-1000 is used to measure volumes between 100 and 1000 μL accurately. The P-200 is used to measure 20–200 μL. The P-20 is used to measure 1–20 μL.

Always select the micropipette suitable for the volume you need to measure as they do not work well beyond their designated ranges.

Select one of the micropipettes available and note the range in which it works. The volumes can be adjusted using the knob at the top being careful not to force it out of range. Choose the proper tip size; unless instructed otherwise, blue tips go with the P1000, and yellow tips go with the P-200 and P-20. If the tips are in a box, bring your micropipette down directly on the tip to put it on. To make sure it is secure, push it on the pipette from the wide end of the tip. Try not to touch the tip that will go into your lab liquids! Grip the pipette in your dominant hand and use your thumb to push on the plunger. Note that there are two "stops", first a "soft stop" and then, if you keep pushing, a "hard stop". To draw fluid into the pipette, push down to the "soft stop", insert the tip into the fluid, and, with your thumb on the plunger, slowly draw fluid into the tip as the plunger comes back up. Look to make sure there are no bubbles. To deliver fluid, place your tip wherever you want the fluid to go and push all the way down to the "hard stop". Withdraw the pipette while holding down on the plunger (why?). Remove the tip using the ejection button (or by grabbing the base) and dispose it in the labeled tip waste container.

Now try it! We will transfer a series of volumes from your tube of blue fluid and deliver that fluid to a piece of parafilm on the bench top.

Exercise 1.18

(a) Using the micropipette that you practiced with, set a volume of your choice, and draw up that volume of water or dyed water into the appropriate tip. Note how full it is and draw the filled tip in your lab notebook. Now eject the volume onto a parafilm square. Draw the size of the drop in your notebook. Label your drawings with the volume you measured.

(b) Repeat using the following volumes measured with an appropriate micropipette and tip: 1 mL (1000 μL), 0.5 mL (500 μL), 0.2 mL (200 μL), 0.1 mL (100 μL), 0.02 mL (20 μL), 0.01 ml (10 μL), 0.002 mL (2 μL).

Serial dilution with dye solutions and water

We will prepare 1 mL serial dilutions to examine "by eye" and with the spectrophotometer using a stock dye solution and distilled water. You will serially dilute your dye solution 1:1 with deionized water. First, label five test tubes near the top 1, 2, 3, and so on using the labeling tools provided.

The next step is to add 1 mL of the deionized water to each tube using the appropriate pipette and tip.

Begin the serial dilution by transferring 1 mL of the dye solution into tube 1 containing 1 mL of water. Dispose of your tip. Mix the dye and water thoroughly by vortexing tube 1 for 3 seconds or capping and inverting it two or three times.

Now transfer 1 ml from tube 1 into tube 2 and mix the second tube thoroughly. Continue this process remembering to mix after each transfer. Note that your last tube should have 2 ml of fluid.

Observe the intensity of the colors in your tubes. Describe what you see in your notebook. Keep these tubes for the next step, in which we will measure the relative concentration of dye as the absorbance of each of your dilutions.

Measuring absorbance

Turn on the spectrophotometer and follow specific instructions for setting the wavelength (e.g., 410 nm for blue food coloring). Set up your cuvettes in a cuvette rack and label them at the very top, for example, B for "blank", 0 for undiluted dye, and 1, 2, 3, and so on for your serial dilutions. Remember to handle them only at the top or by holding the edges so that you do not smudge the polished sides.

The first step is to either read the value of the "blank" or to set the zero absorbance value for the spectrophotometer. In this case, our "blank" is simply deionized water without the dye. Put 1 mL of water into a cuvette and place it as instructed into the cuvette holder; the direction matters to make sure the light is passing through the clear polished sides of the cuvettes. Close the lid and "zero" the sample by pressing 0 Abs. It should read 0.000 Abs.

Pipette 1 mL of your undiluted dye into another cuvette (the one labeled "0" indicating no dilution) and place it in the spectrophotometer. Record the absorbance value in your notebook.

Repeat this procedure for each of the diluted samples taking care to record the absorbance of each.

Graphing and data analysis

These exercises can easily be done in your lab notebook, but you are welcome to use R if you prefer.

Exercise 1.19

(a) Plot the recorded absorbance values on the y-axis versus the tube or dilution number from 0 (undiluted dye solution) to 5 or 6. What is the shape of the curve?

(b) Each of the dilution steps was 1:1, so that the new solution has a concentration of 1/2 that of the solution before it. Do your Absorbance values go down by 2 for each subsequent dilution? What would you expect would happen if you had diluted in steps of 1:4 (1 mL dye to 4 mL water)? What type of mathematical relationship is this?

(c) We said that one of the beauties of absorbance measurements is that they are directly proportional to the concentration of the solution. Can we see if this is true? First, calculate a relative concentration for each of your samples by assigning the concentration of undiluted dye to be "1" dye unit/mL. (Recall that concentration is an amount per volume.)

(d) Plot the absorbance values on the y-axis against your calculated concentrations on the x-axis. What do you see? Can you fit the points with a straight line? Either estimate from the graph directly or by fitting a line to the data using linear regression. What is the absorbance for a sample that is 0.7 dye units/mL? or 0.1 dye units/mL?

Now that we have completed our measurements and data analysis, be sure to clean up your lab space as directed by your instructor. All used tips and liquids should be in their respective waste containers.

UNIT 2

Introduction to modeling using difference equations

Learning outcomes

- Learn and engage in all steps of the modeling process.
- Formulate discrete-time equations that model a range of biological problems and identify assumptions of each model.
- Practice stating assumptions, asking specific questions, using units, and interpreting outcomes.
- Estimate parameters using linear regression.
- Gain confidence in using R.

To begin our exploration of using mathematical models to study biological phenomena, we will use discrete-time equations to model a variety of biological phenomena and linear regression to estimate their parameters. These relatively simple mathematical approaches are extremely useful for dynamic systems that change in predictable ways, especially if we can find a linear relationship to describe the phenomena. We will use these approaches to address a variety of questions surrounding population and community dynamics and the flux of a simple drug through our bodies. We will begin to learn how to use mathematical equations to represent changes in some variable(s) and then use data sets (existing or obtained in the lab) to extract model parameters that will help us to understand how a system works, make predictions, and/or control the system to get a particular outcome. Specifically, we will examine a number of population growth rate models for recovering whooping crane populations and bacteria under a variety of constraints. We will look at more complicated scenarios when populations oscillate (garlic mustard) due to limitations to growth and intraspecific feedback controls on "birth" and "death" rates. We will explore how communities of species assemble on an island by determining colonization and extinction rate parameters, estimating the equilibrium number of species, and exploring the relationships between the rate constants and island size or distance from a population source using the classic island biogeography theory of MacArthur and Wilson. Not all of our models involve populations, but we can use the same tools to model other variables such as the concentration of a drug in our bodies over time. We will determine the rates of uptake and elimination rates for caffeine, from which we can determine when and what the maximum concentration will be. In all cases, once we

Exploring Mathematical Modeling in Biology Through Case Studies and Experimental Activities
https://doi.org/10.1016/B978-0-12-819595-6.00008-6

have the basic parameters, we can explore how different preconditions or assumptions will change our results (e.g., what is the effect of body weight on the maximum concentration of caffeine?).

To begin, we will introduce the mathematics and specific R code we need with some examples. This will be followed by three longer case studies, which will challenge us to dive more deeply into a problem and address interesting questions with our models. Finally, we will have an opportunity to combine our ability to design and carry out experiments to answer questions using our parameter estimation skills applied to our data and a logistic growth model.

2.1 Discrete-time models

In this unit, we consider phenomena for which it is reasonable to consider changes in the underlying variables at discrete time steps. For example, suppose we used plate counts to count the number of living bacteria on agar plates each hour for eight hours. This would result in a sequence of 9 data points for the independent variable (time) and dependent variable (population). It might be reasonable to model how the population changes each hour instead of continuously in time. Another example is to consider a population whose nesting season is in the spring. If we collect data annually before eggs hatch, then we could use a discrete model to study how the population changes from year to year.

We can denote a population as the sequence of numbers $N_0, N_1, \ldots, N_k \ldots$, where N_0 is the initial population, and N_k is the population after k fixed time steps. The time step chosen depends on the particular phenomena being modeled. Although discrete-time models assume that events take place at discrete intervals (e.g., an insect that breeds synchronously in the fall), continuously breeding populations can sometimes be represented adequately using a discrete-time model. In this unit, we introduce discrete-time models as an intuitive approach to setting up models, but in later units, we will develop models in which time is assumed to be continuous.

Consider a generic example in which the population in year k, denoted by N_k, depends on the population at previous times. We refer to this as a *difference equation*. The simplest assumption is that the population depends only on the population at the previous time step (a *first-order difference equation*), that is,

$$N_{k+1} = f(N_k), \quad k = 0, 1, \ldots. \tag{2.1}$$

The function $f(N_k)$ may be a simple linear function or a more complicated nonlinear function (as you will see in the Invasive Species Case Study!). For now, it is best to start as simple as possible. Often we can isolate N_k as a separate term on the right-hand side:

$$N_{k+1} = N_k + F(N_k), \tag{2.2}$$

where the right-hand side represents the population at the previous time step plus the change in population in one time step. In general, the number of individuals in a

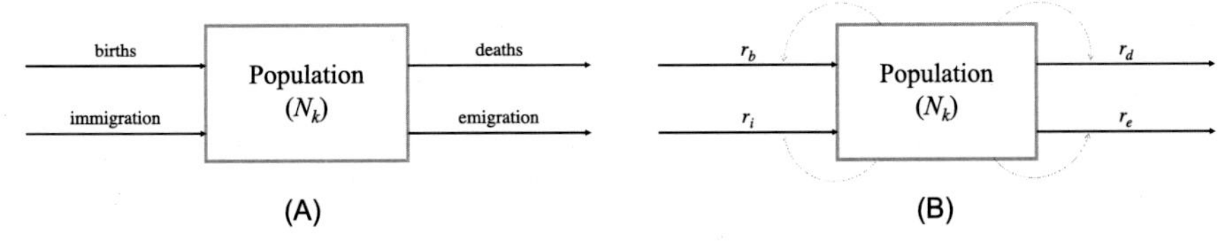

FIGURE 2.1

(A) Qualitative diagram of how a population can change in each time step. (B) Qualitative diagram that includes rate constants and influences (dashed curves).

population can change by four processes: births, deaths, immigration, and emigration. These inflows and outflows can be represented in a qualitative diagram (Fig. 2.1A).

Mathematically, there are many ways we might express these flows. Assuming on average each individual gives birth to the same number r_b of offspring each time step, the number of births in year k will be $r_b N_k$. The probability that an individual will die is denoted by r_d, so that the number of deaths in year k is $r_d N_k$. The parameters r_b and r_d are called the per capita birth rate and per capita death rate, respectively. We might also assume the number of immigrants and emigrants each time step are proportional to the population, that is, $r_i N_k$ and $r_e N_k$. In general, the parameters r_b, r_d, r_i, and r_e are called *rate constants* since their units are 1/time. The qualitative diagram can be modified to include the rate constants, along with dashed arrows that represent the variables influencing each rate (see Fig. 2.1B). We can now write the model as a mathematical equation

$$N_{k+1} = N_k + r_b N_k + r_i N_k - r_d N_k - r_e N_k,$$

which can be written as

$$N_{k+1} = (1 + r)N_k \tag{2.3}$$

where $r = r_b + r_i - r_d - r_e$. Rearranging Eq. (2.3), we have

$$\frac{N_{k+1} - N_k}{N_k} = r. \tag{2.4}$$

We see that r is equal to the change in population divided by the population, and therefore r is called the per capita growth rate. Note that in this model the per capita growth rate is assumed to be constant. In the lab, we will consider a more complicated per capita growth rate, which depends on the population density.

Exercise 2.1

Assuming a population can be modeled by Eq. (2.3), for what values of r would the population increase over time? For what values of r would the population decrease? For what values of r would the population remain constant? What do these results imply about the relationship among the specific rates for births, deaths, immigration, and emigration?

Every model is a simplification of the underlying biology, and it is important to be explicit about these assumptions when formulating a model. Since we are using a difference equation, we are assuming that no processes occur between increments of time. Can you articulate other assumptions of our model? When might they be reasonable? We will explore these types of questions in the case studies and lab.

2.1.1 Solutions to first-order difference equations

A *solution* to a difference equation $N_{k+1} = f(N_k)$ is the value of N over time, that is , N_0, N_1, and so on. Computing the solution iteratively, which is typically done using a computer, is an example of a *numerical solution*. As an example, consider the model $N_{k+1} = (1+r)N_k$. If the initial population is 3 and the per capita growth rate r is 1 per year, then the model becomes $N_{k+1} = 2N_k$ with initial condition $N_0 = 3$. We can iterate to find $N_1 = 2N_0 = 2 \cdot 3 = 6$, $N_2 = 2N_1 = 2 \cdot 6 = 12$, and we can continue in this manner to compute the population at future times. R can quickly iterate this recursive equation using a `for loop` as introduced in Section 1.2.6. See Listing 2.1 for an example.

Iterating Difference Equations

Loops are particularly useful for iterating difference equations. Consider the difference equation $N_{k+1} = 2N_k$ with $N_0 = 3$. We can calculate $N_0, N_1, \ldots, N_{15}$ using the commands in Listing 2.1.

Listing 2.1 (unit2ex1.R)

```
# Create vector to store N0, N1,..., N15 and store initial value
N=rep(0,16); N[1]=3;

# Iterate to compute N_k
for (i in 1:15) {
  N[i+1]=2*N[i]
}
plot(0:(length(N)-1),N,xlab="time step",ylab="population")
```

As an exercise, change the `for loop` to be `for (i in 1:3)`. Work through the code by hand and write the output for each time through the loop ($i = 1, 2, 3$) and then plot the output by hand. Run this in R and compare to the steps you performed on paper. This can help you become more comfortable with setting up loops!
Note: Since indexing in R begins at 1 (just a rule), N_0 in our model formulation is equivalent to `N[1]` in R!

Exercise 2.2

Not all discrete-time models are as simple as the previous example, yet these can still be solved using iteration (for loops) in R. Consider the difference equation $N_{k+1} = N_k e^{-0.01 N_k}$ with $N_0 = 10$.

(a) Calculate N_1 and N_2 by hand.
(b) Write a script file that uses a `for loop` to compute $N_0, N_1, \ldots, N_{10}$. Plot N_k versus k.

Exercise 2.3

Consider the difference equations

$$p_{k+1} = p_k + 0.1 p_k, \tag{2.5}$$

$$q_{k+1} = q_k + 0.1 q_k \left(1 - \frac{q_k}{30}\right), \tag{2.6}$$

where p and q represent two different populations, and k represents the number of days.

(a) We can rearrange Eqs. (2.5) and (2.6) to explicitly show the change in each time step:

$$\begin{aligned} p_{k+1} - p_k &= 0.1 p_k, \\ q_{k+1} - q_k &= 0.1 q_k \left(1 - \frac{q_k}{30}\right). \end{aligned}$$

Just by inspection of the equations, compare the expressions for the changes in these two populations in each time step. For each model, is the population always increasing, decreasing, or does it depend on the population size?

(b) Assume that both starting population sizes are the same, specifically, $p_0 = 10$ and $q_0 = 10$. Write a script file that uses a `for loop` to iterate Eqs. (2.5) and (2.6) to compute the number of individuals over the first 30 days and plot these resulting two populations p_k and q_k versus k (i.e., day) on the same graph. Discuss the differences in long-term behavior.

(c) Does the initial value for q_0 matter? Iterate the equation for q_k using initial conditions $q_0 = 10$, $q_0 = 20$, $q_0 = 30$, $q_0 = 40$, and $q_0 = 50$. What is the long-term behavior of the population for various initial conditions? Discuss your observations. What happens when $q_0 = 30$? In this example, the 30 in the equation represents a population term called the carrying capacity (often denoted K) or the maximum number of individuals that can be sustained over the long term. We will explore this model further in the lab!

Although it is always possible to iterate a discrete difference equation to determine the population at future times, it can be helpful to instead use the *analytical solution*, which is an explicit solution for the variable at any time. For example, given a difference equation of the form $N_{k+1} = f(N_k)$, the explicit solution could be written in the form $N_k = g(k)$, where the population is written as a function of the time step. Having an explicit solution means that we do not have to iterate the difference equation to compute the population at intermediate times. Whereas it is convenient to have an analytical solution, it is only possible to obtain these closed-form solutions for very simple difference equations. For example, the difference equation $N_{k+1} = (1 + r)N_k$ proposed earlier is simple enough that the analytical solution can be quickly derived by repeatedly substituting previously computed values of the population as follows:

$$\begin{aligned} N_1 &= (1 + r)N_0 \\ N_2 &= (1 + r)N_1 = (1 + r)(1 + r)N_0 = (1 + r)^2 N_0 \end{aligned}$$

$$\begin{aligned} N_3 &= (1+r)N_2 = (1+r)(1+r)^2 N_0 = (1+r)^3 N_0 \\ &\vdots \\ N_k &= (1+r)N_{k-1} = (1+r)^k N_0. \end{aligned} \tag{2.7}$$

Therefore, given N_0 and r, we can use $N_k = N_0(1+r)^k$ to plug in any k and calculate N_k. We can check that $N_k = (1+r)^k N_0$ is a solution to $N_{k+1} = (1+r)N_k$ by plugging the solution into the left-hand side N_{k+1} and right-hand side $(1+r)N_k$ and verifying that they are equal:

Left-hand side: $N_{k+1} = (1+r)^{k+1} N_0$.

Right-hand side: $(1+r)N_k = (1+r)(1+r)^k N_0 = (1+r)^{k+1} N_0$.

We see that the left-hand side is equal to the right-hand side, and therefore $N_k = (1+r)^k N_0$ is in fact a solution to the difference equation $N_{k+1} = (1+r)N_k$. This is lucky! Many of the difference equations used to model biological systems are more complicated and have no analytical solutions, and therefore we may need to use other methods to study the behavior of solutions. See [15] for a thorough discussion of difference equations.

Exercise 2.4

Consider the model $N_{k+1} = 1.1N_k$ where $N_0 = 2$.

(a) Use a `for loop` in R to compute N_k for $k = 0, \ldots, 30$ and plot N_k versus k.

(b) Now use the explicit solution $N_k = 2 \cdot 1.1^k$ to compute N_k at $k = 0, \ldots, 30$ and plot N_k versus k. Note that you do not need to use a `for loop` to do this! Instead, create a vector k and evaluate $2 \cdot 1.1^k$ at these values of k. Is your result the same as part (a)?

2.1.2 Using linear regression to estimate parameters

Mathematical models include parameters that must be assigned values. For example, given $N_{k+1} = (1+r)N_k$, the parameter r must be determined before we can think about projecting population size. This process is called *parameterization*. Parameter values can sometimes be found by performing specific experiments or by literature searches. However, often parameter values are found by *calibration*, where parameters are estimated by fitting the model to data. In this unit, we will focus on models that can utilize linear regression for fitting: we will explore other methods of calibration in later units.

Consider the example data set of population density (number per area) versus time (year), given in Fig. 2.2. By inspection the population looks as though it could be growing exponentially. From the previous section, we know that the solution to $N_{k+1} = (1+r)N_k$ is an exponential model, and therefore this model might capture the dynamics of this system.

k	Population density in year k
0	2.19
1	2.39
2	2.86
3	3.23
4	3.11
⋮	⋮
17	10.46
18	10.57
19	12.72
20	13.04

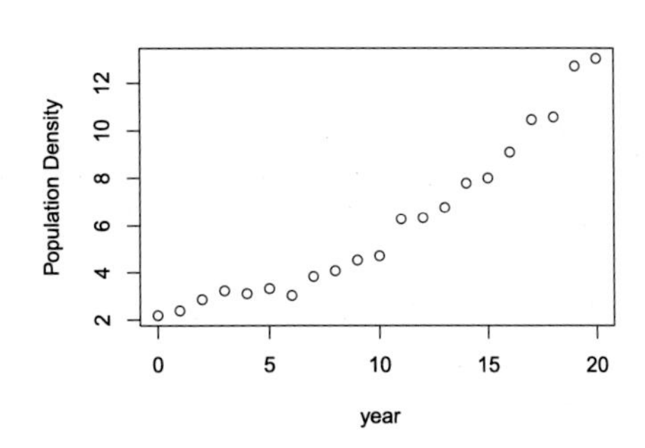

FIGURE 2.2

Table of simulated annual population density over twenty years. Density versus time is plotted.

k	Population density in year k	Population density in year $k+1$
0	2.19	2.39
1	2.39	2.86
2	2.86	3.23
3	3.23	3.11
4	3.11	3.32
⋮	⋮	⋮
17	10.46	10.57
18	10.57	12.72
19	12.72	13.04
20	13.04	

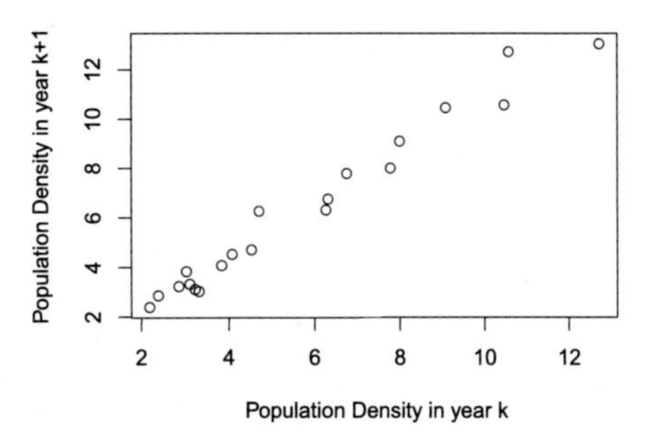

FIGURE 2.3

Table of simulated annual population density and the subsequent year's population density. Density data are plotted as a function of the density in the previous year.

Our model relates N_{k+1} to N_k, but the data give population versus time. Can we transform the data to see if this model might be a good fit? Note that the model $N_{k+1} = (1 + r)N_k$ assumes a linear relationship between N_{k+1} and N_k with slope $1 + r$ and vertical intercept 0. We already have data for population N_k in year k, and for each year, we can write the corresponding population N_{k+1} in the following year. We can create this new vector as shown in Fig. 2.3 and plot the population in year $k + 1$ versus the population in year k. Aha! The relationship looks approximately linear, which gives us confidence in our model.

To calibrate our model, we want to find the value of r that gives the best match between the model and data. In Section 1.2.7, we learned how to run a linear regression in R. Here, the model $N_{k+1} = (1 + r)N_k$ assumes a linear relationship between N_{k+1} and N_k with slope $1 + r$ and vertical intercept 0. If we perform linear regression on the data shown in Fig. 2.3, then we can find the slope, given by $1 + r$, and then calculate r. Details of the data manipulation and linear regression analysis are shown in Listing 2.2.

Model Calibration in R: Linear Regression

To find the best-fit parameter r in $N_{k+1} = (1+r)N_k$, we must first form a vector N_k and the corresponding vector N_{k+1} before performing linear regression.

Listing 2.2 (unit2ex2.R)

```
# Input data. k in years and N in number per unit area
k = seq(0,20)
N = c(2.19,2.39,2.86,3.23,3.11,3.32, 3.03,3.83,4.07,4.52,4.70,6.26,6.31,
           6.75,7.78,8.00,9.09,10.46,10.57,12.72,13.04)

# Form density vectors for year k and k+1 as shown in Fig. 1.3.
# Note that these need to be the same length!
n = length(N)
N_k = N[1:n-1]
N_kplus1 = N[2:n]

# Plot N_(k+1) versus N_k (Fig. 1.3)
plot(N_k,N_kplus1,xlab="Population Density in year k",
     ylab="Population Density in year k+1")

# Perform linear regression and plot line
fit=lm(N_kplus1~N_k+0) # Use +0 to force line to go through origin
abline(fit)

# Compute r from linear regression output
r = fit$coefficients - 1 # slope=1+r, so r=slope-1

# Plot data (N versus k) and overlay model output
plot(k,N,xlab="year",ylab="Population density") # Plot data
N0 = N[1]  # Set N_0 equal to first data point
points(k,N0*(1+r)^k,pch=20)  # Plot model solution using the estimated r value
legend("topleft",c("data","model"),pch=c(21,20))
```

Read carefully through this listing, run it in R, and make sure you understand the purpose and output of each line.

Exercise 2.5

The pine looper moth is a dangerous forest pest in parts of western United States, western Canada, and northern Europe. Females lay their eggs on pine or other conifer needles. When the larvae hatch, they feed on both old and new needles. Defoliation can lead to tree mortality, especially in younger trees, and although mature trees may survive the winter after defoliation, they often succumb to secondary attack by bark beetles. Pine looper moth pupae overwinter on the forest floor. Data from a study conducted in Cannock Chase Forest in England between 1960 and 1970 are given in Table 2.1.

Table 2.1 Density of the pine looper moth pupae in Cannock Chase Forest [7].

Year	Density (no. pupae/m^2)
1960	20.31
1961	10.81
1962	11.33
1963	1.72
1964	0.45
1965	0.29
1966	2.04
1967	10.24
1968	7.94
1969	2.01

Pine looper moths have nonoverlapping generations, and therefore a discrete-time model is suitable to model this population. Here we will fit Ricker's model to the data. This model was developed in 1954 to study the salmon population in the Pacific northwest and, similar to the logistic growth model, assumes a carrying capacity. However, unlike the logistic growth model, Ricker's model does not allow the population variable to become negative (which can happen in the logistic growth model if the population is large!). Ricker's model is given by

$$N_{t+1} = N_t e^{r\left(1-\frac{N_t}{K}\right)}, \tag{2.8}$$

where N_t is the density of pupae at time step t, r is interpreted as the intrinsic growth rate, and K is the carrying capacity or maximum density of individuals that can be sustained in the area.

(a) Explain why K is interpreted as the carrying capacity in the model given in Eq. (2.8). (*Hint*: Consider the expression $e^{r\left(1-\frac{N_t}{K}\right)}$ for the cases $N_t = K$, $N_t < K$, and $N_t > K$.)

(b) If we can transform Eq. (2.8) into a linear relationship, then we can use the data and apply linear regression to estimate the parameters r and K. Show that if we divide both sides of the Ricker equation by N_t and take the natural log of both sides, we obtain the linear equation

$$y = mx + b$$

where $y = \ln\left(\frac{N_{t+1}}{N_t}\right)$, $x = N_t$, $m = -\frac{r}{K}$, and $b = r$.

(c) In R, input the data from Table 2.1 (or import the data set MothData.csv) and perform linear regression. (Note that you will need to create vectors for N_t and N_{t+1} from the data, and then create a vector for $\ln\left(N_{t+1}/N_t\right)$.) Report the slope m and intercept b found from linear regression.

(d) Use the slope and intercept found in part (c) to compute r and K.

(e) Plot the data $\ln\left(N_{t+1}/N_t\right)$ versus N_t and add the regression line to the graph. What are your conclusions about applicability of the model to this data set?

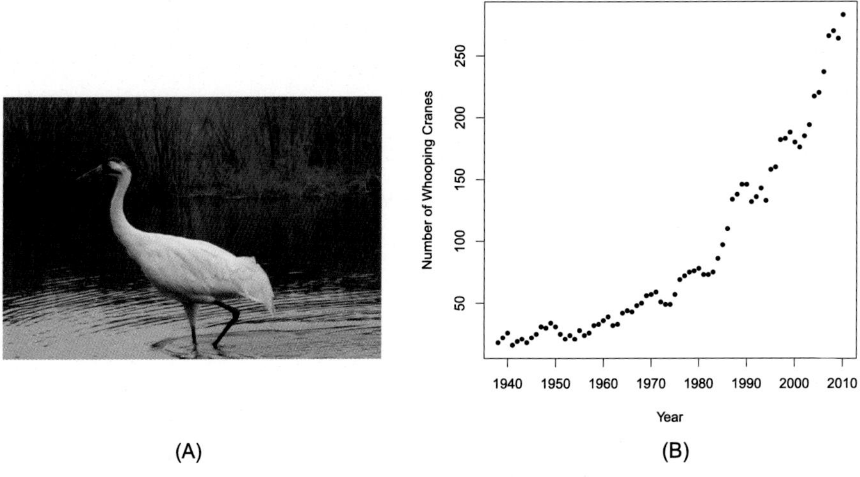

FIGURE 2.4

(A) Adult whooping crane (By CheepShot – Whooping Crane, CC BY 2.0, https://commons.wikimedia.org/w/index.php?curid=37127745). (B) Whooping crane population on the wintering grounds surrounding Aransas National Wildlife Refuge, Texas, USA (Data from Table 2 in [8]).

2.2 Putting it all together: the whooping crane

Now that we have begun to build our modeling toolkit, we will consider data on a whooping crane population and step through the modeling process as outlined in Fig. 1.1. Whooping cranes were nearly extinct in the 1940s when their population fell to less than 20 individuals. Widespread habitat change along with unregulated shooting are thought to have caused the long-term population decline. Due to Endangered Species Act protection, the population is on the path to recovery, as observed in Fig. 2.4B, where each data point represents an annual abundance count of whooping cranes, which is typically lower than the total population but still a good representative of the population trend.

Exercise 2.6

Import the data Cranedata.csv into RStudio. Plot the number of whooping cranes over time, replicating Fig. 2.4B.

The only remaining whooping crane natural population nests in Wood Buffalo National Park in Canada and spends the winter at the Aransas National Wildlife Refuge on the Texas gulf coast. Cranes are known to live at least 22 years in the wild (perhaps as long as 40 years) and produce their first fertile eggs between ages 4 and 7. Whooping cranes typically rear only one offspring per year.

Objective. The whooping crane is currently listed as federally endangered. According to the International Whooping Crane Recovery Plan, the Aransas-Wood

Buffalo population must consist of 1000 individuals and 250 reproductive pairs before the species can move from the endangered list to the threatened list. Before thinking of the complications in modeling the substructure of the population (males, females, reproducing) we can develop a simple model to predict the population growth assuming current conditions and the expected time to reach the goal of 1000 individuals.

Setting up the model. We can denote the population as the sequence of numbers $N_0, N_1, \ldots, N_k$ where N_k represents the number of whooping cranes k years after 1938. Note that the initial data point in Fig. 2.4B is in 1938 and is given by $N_0 = 18$. We would like to write the population in year k in terms of the population in the previous years. Assuming that the population depends only on the population in the previous year, we have

$$N_{k+1} = N_k + f(N_k), \quad k = 0, 1, \ldots. \tag{2.9}$$

Recall that in general the number of individuals in a population can change by four processes: births, deaths, immigration, and emigration. However, since there is only one wild population, we will only consider births and deaths. In other words, we have

$$\begin{aligned}&(\text{\# of whooping cranes in year } k+1)\\&\quad = (\text{\# of whooping cranes in year } k) + \text{births} - \text{deaths}.\end{aligned}$$

Since we do not have explicit birth and death data (only the population abundance), we will begin with the assumption that the numbers of births and deaths are proportional to the population, which gives

$$N_{k+1} = (1 + r)N_k, \tag{2.10}$$

where $r = r_b - r_d$ is the per capita growth rate.

Parameter estimation. In Eq. (2.10), r is an unknown parameter. Given the data on population size of whooping cranes on the wintering grounds at Aransas National Wildlife refuge, we can fit the model in Eq. (2.10) to the data.

Exercise 2.7

Use linear regression to estimate r and reproduce Fig. 2.5A. Refer to Section 2.1.2 for guidance. Note that the difference equation $N_{k+1} = (1 + r)N_k$ assumes a linear relationship between N_{k+1} and N_k, where $(1+r)$ is the slope, and the y-intercept is zero. Therefore the line shown in Fig. 2.5A is constrained to go through the origin.

Model predictions. We can visualize the data and model output in another way that might be more meaningful, population versus time (see Fig. 2.5B).

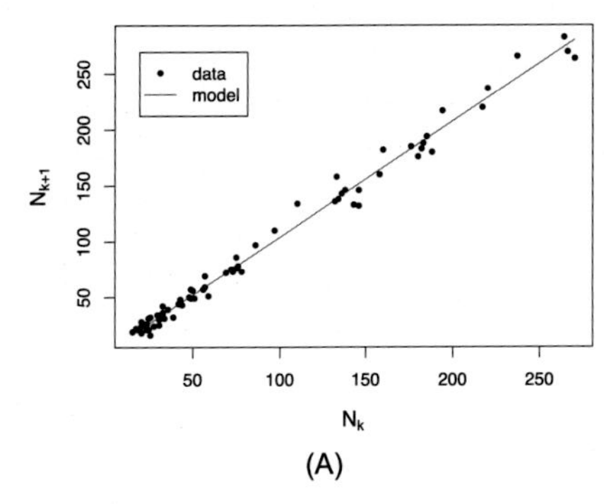

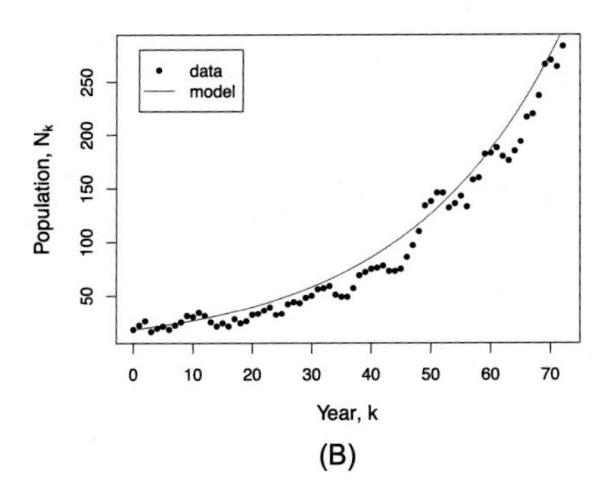

FIGURE 2.5

(A) Representation of data as N_{k+1} vs. N_k and best-fit line estimated by forcing the line through 0,0. (B) Time series data and model output.

Exercise 2.8

Compare the model output, that is, the number of whooping cranes over time, to the data. Produce a figure that looks like Fig. 2.5B. Recall that there are two ways to generate the output of this model. You can use a `for loop` to iterate Eq. (2.10) and compute the population each year. Alternatively, you can use the explicit solution $N_k = N_0(1+r)^k$.

Recall that the objective of the whooping crane model was to predict when the population will reach 1,000 cranes, assuming that the population continues to grow along its current trajectory (that is, assuming that the parameter r is constant over time). We can iterate Eq. (2.10) until the population reaches 1,000, or we can use the explicit solution $N_k = N_0(1+r)^k$ to solve for the unknown time k exactly! Rearranging $N_k = N_0(1+r)^k$, we find

$$k = \frac{\ln \frac{N_k}{N_0}}{\ln(1+r)}, \tag{2.11}$$

which provides a formula for the number k of years it takes to reach a population N_k.

Exercise 2.9

Using your estimate for r and initial condition $N_0 = 18$, calculate the year in which the crane population will reach 1000 predicted by the model.

Further questions. Our proposed model assumes a constant per capita growth rate $r = (N_{k+1} - N_k)/(N_K)$, which results in exponential growth. This can be a good approximation initially, but eventually there are environmental and demographic effects that might limit growth as alluded to in problems with an explicit carrying capacity. This will be discussed more in Lab 1.

The objective of our model was to forecast future growth, but there are other questions we can explore. We might modify our model to better understand how specific parameters affect whooping crane population dynamics. For example, drought conditions may directly increase mortality of whooping cranes on their winter territories. Butler et al. [9] modeled winter loss and its contribution to annual mortality as functions of winter drought, which may explain some of the oscillations in the actual population data. Understanding these effects can guide management actions that may mitigate winter mortality.

2.3 Case study 1: Island biogeography

Goals: We will formulate a discrete-time model to address whether the number of species on an island reaches an equilibrium based on immigration and extinction, and explore how these rates vary with physical factors such as the island size or distance from a source of new species.

2.3.1 Background

Island biogeography is an important concept in ecology and evolutionary biology due to the out-sized contributions island species make to overall biodiversity (3.5% of total land area and 15–20% of total species) and the fact that they are small enough to be sampled thoroughly [73]. MacArthur and Wilson published their paradigm shifting monograph *The Theory of Island Biogeography* in 1967 outlining a simple model designed to describe the accumulation rate of species on an island as a function of immigration and extinction rates. They propose that these parameters, in turn, vary with the number of species in a given taxon present on a given island. The number of species will also depend on physical factors, for example, the distance (difficulty of immigration) and size (capacity for species) of the island of interest. Their theory predicts an equilibrium number of species for a given island that will be achieved when immigration and extinction rates are equal. Their prediction was based on a large data set showing the number of species on an island directly correlates with its size (explored in Exercise 1.16). This is similar to the idea of the carrying capacity K of a single population in a given area (as explored in the lab) but applied to the total number of species.

The importance of this model lies in its ability to make testable hypotheses comparing different types of organisms or to explore more complex reasons for species variability such as evolution (Darwin's finches) or the number of available niches for new species. Hundreds of papers have been written on all sorts of "islands," literal ones such as Hawaii, isolated areas such as mountaintops, bogs, forest clearings, lakes, and even individual plants. Many scientists use island biogeography models to project or evaluate conservation strategies such as the design of nature reserves [75] or to provide a framework for the emerging field of landscape ecology.

One of the challenges of testing the MacArthur and Wilson model is that colonizing a remote island takes a long time, possibly millions of years, which makes obtaining data sets particularly challenging. Fortunately, there are data from new islands (islands in Thousand Island Lake, which formed after the Xin'an Dam was built in China), islands that suffered a disaster such as a volcano that killed all existing species (e.g., Rakata Island), forest clearings, or, over longer time, immigration and extinction rate estimations using molecular clocks (e.g., New Zealand ferns and lycophytes). Despite the data collection challenges, the MacArthur and Wilson model is still being applied more than 50 years after its publication, with many studies focused on the concept of species number equilibria and others on the determinants of colonization rates.

The island biogeography model suggests that immigration and extinction rates for species determine the number of species on an island at any given time. Immigration is assumed to come from a source population(s) some distance from the island, and the rate of new species immigration is expected to decrease with increasing distance or isolation and with the number of species already on the island. The extinction rate is hypothesized to be inversely related to island size and directly to the number of species. Finally, the number of species in a given group or taxon of organisms is expected to reach equilibrium.

Applying this model raises many questions. For example:

1. What is the source population for immigration?
2. Why will distance and existing species numbers alter immigration rate?
3. Why will area of the island and the existing number of species alter the extinction rate?
4. What other factors might affect the number of species, the immigration, or extinction rates?
5. Why might we consider species by taxon or similar species rather than all possible species?
6. Why might it be useful to know how many species a given island could hold?

After addressing these questions, it is clear that the MacArthur and Wilson model is simple with a well-defined framework. The critical concepts in this framework include the idea that successful immigration (versus just arrival) assumes that the number of individuals arriving is enough to propagate and that there is food and habitat available to do so. The latter factors depend on what species are already present and the geography of the island itself. To remind us that we are considering new species becoming established on the island, we will switch our language at this point from immigration to colonization rate. Second, the number of new species to arrive will decrease with the number present already (decreased probability of being new) and extinction will depend on the number present to go extinct and the competition among existing species. Finally, it is important to note that for the island biogeography model, when we say "the number of species," we mean of a given taxon or type of organism. It does not take much imagination to realize that the immigration rate for birds is likely to be quite different than that for wind-borne plant seeds and that the niches required for each are different.

How can we translate these ideas into a mathematical model?

2.3.2 Model formulation

MacArthur and Wilson [41] originally developed their theory of island biogeography to explain the species-area relationship, that is, the relationship between the area of a habitat and the number of species found within that area (see Exercise 1.12). They believed there could be a balance of colonization and extinction leading to an equilibrium that could explain the diversity on islands. Developing a model that shows how the number of species changes due to colonization and extinction rates can help us estimate the number of species at equilibrium and if that equilibrium has been attained. A conceptual model of how the number of species S_n changes per time step is depicted in the following diagram:

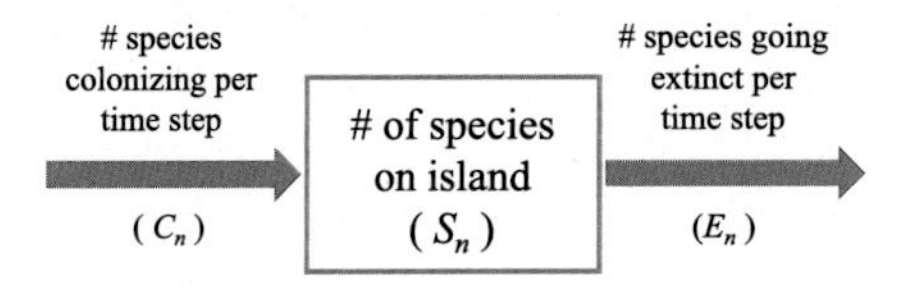

We can translate this diagram to mathematical form:

$$S_{n+1} = S_n + C_n - E_n. \tag{2.12}$$

Rearranging this equation to $S_{n+1} - S_n = C_n - E_n$, the left side represents the change in the number of species per time step, which is equal to the colonization rate minus the extinction rate. Writing the equation in this form helps us see that the units of C_n and E_n are (number of species)/time, since the units of these terms must be equivalent to the units of the left side, $S_{n+1} - S_n$, change in species over a single time step from n to $n + 1$.

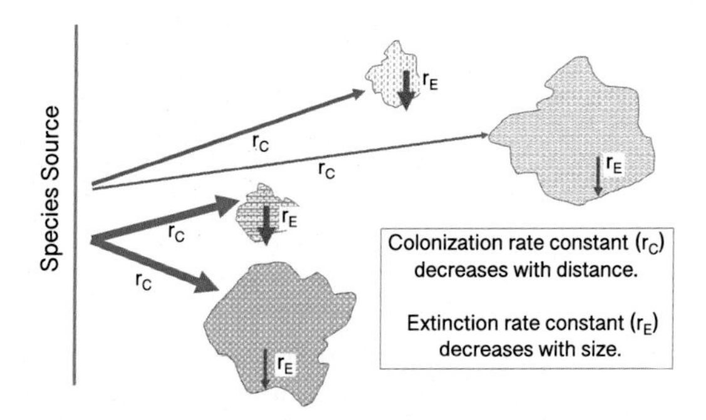

FIGURE 2.6

Islands are colonized by species immigrating from a larger landmass with rate constants (horizontal arrows) that vary with distance or difficulty of transport. They lose species by extinction with rate constants (downward arrows) inversely proportional to their size.

Specifically, MacArthur and Wilson assume the following about the colonization and extinction rates as suggested in Fig. 2.6:

1. The colonization rate decreases as the number of species on the island increases, and the colonization rate will be zero if all species from the mainland source (denote this number by P) have colonized the island.
2. The extinction rate on the island increases as the number of species on the island increases.

Exercise 2.10

(a) Explain why MacArthur and Wilson's two assumptions might be reasonable.
(b) Using these two assumptions, sketch a possible relationship between the colonization rate C_n and the number of species S_n (put C_n on the vertical axis). On the same plot, sketch the extinction rate E_n versus the number of species S_n.

Exercise 2.11

When formulating a mathematical model, it is a good practice to start as simple as possible! The simplest relationships we can assume between C_n and S_n and between E_n and S_n are linear relationships. Assume that

$$C_n = r_C(P - S_n), \tag{2.13}$$

$$E_n = r_E S_n. \tag{2.14}$$

(a) Assuming that r_C, r_E, and P are some positive constants, sketch C_n versus S_n and E_n versus S_n on the same plot. Label any intercepts on the horizontal and vertical axes in terms of the model parameters. What are the physical interpretations of these intercepts?
(b) What are the units of r_C and r_E? What are the interpretations of r_C and r_E, both graphically and biologically?
(c) The two lines you drew (C_n and E_n versus S_n) should intersect. What is the interpretation, both mathematically and biologically, of this point of intersection?
(d) Do you think it is reasonable to assume that C and E linearly depend on S? If not, what other relationship would you propose and why?

Substituting the annual colonization and extinction rates in terms of S_n and P into Eq. (2.12) gives the following model that we will explore in this case study:

$$S_{n+1} = S_n + r_C(P - S_n) - r_E S_n. \tag{2.15}$$

2.3.3 Rakata story

The model given in Eq. (2.15) describes changes in the number of species over time and assumes that $t = 0$ corresponds to the origin of the island ($S_0 = 0$). Therefore data from long periods of time are required to obtain estimates of the model parameters. A long-term data set does in fact exist for vascular plant species on the island of Rakata following a volcanic eruption in 1883. In 1883, Rakata was part of a large volcanic island known at the time as Krakatoa. The 1883 eruption was so violent that it ripped the island into four smaller islands and eradicated most life on those islands. Therefore the total number of species on the islands dropped to essentially zero, which is an ideal condition for testing island biogeography models. Soon after the eruption, data were collected and continue being collected through the 1980s. We will use these data to explore how we might estimate parameters in our model and whether or not the model captures the observed dynamics.

Data

Thornton et al. [69] and Whittaker et al. [72] compiled survey data collected since the 1883 eruption to estimate immigration and extinction rates for vascular plants. The data are given in Table 2.2, where S represents the number of species, C is the rate of colonization in species per year, and E is the rate of extinction in species per year.

Table 2.2 Approximate vascular plant data on Rakata given as the number of species S, and the colonization and extinction rates per year for each year specified [72,69].

Year	1883	1886	1908	1920	1933	1986
S	0	24	100	200	220	250
C	8.0	5.0	6.1	4.0	2.3	1.76
E	0.0	0.01	0.5	1.7	1.7	1.7

Exercise 2.12

(a) Create vectors in R for t (years since 1883), S, C, and E.

(b) On separate graphs, plot S versus time in years since 1883 (see Fig. 2.7 for example), C versus time, and E versus time. Describe in words the behavior of each variable with time.

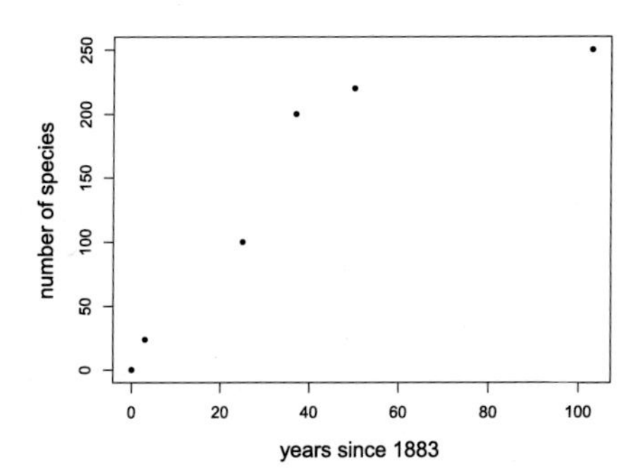

FIGURE 2.7

Number of species over time, where time is years since 1883 (data from [72,69]).

Exercise 2.13

Plot C versus S and E versus S. How do C and E change as S increases? Do these relationships match the general theory provided by MacArthur and Wilson? Explain.

Parameter estimation

The unknown model parameter values are r_C, r_E, and P. Recall, P is the number of species from the mainland source which may be unknown, especially if there are various sources. Once we obtain estimates for these parameters, we can see how well the model captures the dynamics and whether or not the number of species has reached equilibrium.

Exercise 2.14

(a) Recall that we assumed linear relationships between C and S, and E and S. Perform a linear regression analysis to find linear models for both C as a function of S and E as a function of S. (Note: For E versus S we can force the line to go through the origin by using `E~ S+0` in the `lm` command.)

(b) Using `abline` in R, plot these best-fit lines on the same graph, along with the data. How do your results compare to the data?

(c) Use your linear regression results to find estimates for r_C, r_E, and P and report them with units.

Model analysis

Now that we have estimated the model parameters, we can set up the difference equation and iterate the model to compute the model prediction for number of species over time.

Exercise 2.15

Using Listing 2.3 as a template, iterate Eq. (2.15) using the estimates you found for the model parameters. Then plot your model output of S versus time, along with the raw data. Label axes and include a legend. Discuss your observations.

Listing 2.3 (unit2ex3.R)

```
n_end=110; # number of time steps, one time step = 1 year
modS=rep(0,n_end+1)  # Create vector to store values of S
modS[1]=0  # First element of S corresponds to initial number of species
for(i in 1:n_end){
  C = INSERT EXPRESSION HERE
  E = INSERT EXPRESSION HERE
  modS[i+1] = INSERT EXPRESSION HERE
}
INSERT COMMANDS TO PLOT MODEL OUTPUT (NUMBER OF SPECIES VERSUS TIME)
```

One of the benefits of MacArthur and Wilson's model is that it provides a prediction for the number of species after a long time, that is, the equilibrium value for the number of species. What does it mean for our system to be at equilibrium? Equilibrium occurs when the number of species colonizing the island per year is equal to the number of species going extinct each year, that is, the number of species on the island does not change from year to year. In Eq. (2.15), this implies $S_{n+1} = S_n$ or, equivalently, $C_n = E_n$.

Exercise 2.16

(a) Using your plot of S versus t from Exercise 2.15, estimate by eye the equilibrium number of species predicted by the model.

(b) We can calculate the exact equilibrium value predicted by the model. Do this by setting $S_n = \hat{S}$ and $S_{n+1} = \hat{S}$ in Eq. (2.15), where $\hat{S}$ denotes the equilibrium solution. Then, using your parameter values found in Exercise 2.14(c), solve for $\hat{S}$. Is it close to your estimate in part (a)?

Now that you have parameterized your model using data and you understand the long-term behavior of the model, consider some of the applications and limitations to the model:

1. Whereas the model predicts that the number of species smoothly approaches the equilibrium number and remains at that value through time, the data show that in reality the number of species will fluctuate around the equilibrium. Why does this make sense?
2. Since the model predicts the equilibrium number of species, we know that if the number of species existing on an island is less than $\hat{S}$, then we can predict that more species can be accommodated unless there are other critical limitations on the species number. Another reason for the number of species to be less than equilibrium would be recent increases in extinction rates. What sorts of historic data could be used to distinguish between these two situations?
3. Recall that this model is dynamic with new species continuing to arrive and others becoming extinct. How might a conservation biologist use this model to determine the turnover rate of species at equilibrium?

2.3.4 Modern approach: lineage data

In the case of Rakata Island, we were able to directly observe the change in number of species over time following the volcanic eruption. What if field data are not available? For example, many islands are large and have accumulated species over millions of years. Molecular techniques can now be used to estimate the timings of colonization and geographical differentiation from sequence data. For example, Perrie and Brownsey [53] used molecular dating techniques on DNA sequences for pteridophytes (ferns and lycophytes), with each genus represented by one New Zealand species and the most closely related non-New Zealand species. Their divergence estimates will be used here to estimate colonization and extinction rates following New Zealand's separation from the Gondwanian landmass around 80 million years ago.

Exercise 2.17

The cumulative number of lineages since colonization of lineages of pteridophytes on New Zealand began are given in Table 2.3. Plot the data. What are your observations?

Table 2.3 Cumulative number of lineages of pteridophytes (ferns and lycophytes) in New Zealand since time of colonization. Data interpolated from Fig. 14.3 in [38].

Time since colonization (10^6 years)	0	4	8	12	16	20	24	28	32	36
Cumulative no. of lineages	1	4	11	15	18	20	23	24	25	28

Model

Similar to the Rakata example, we assume that the number of lineages (denoted by L here instead of S) changes due to colonization and extinction, that is, $L_{n+1} = L_n + C_n - E_n$. In the Rakata example, the unknown parameters were the colonization and extinction rate constants and the source population from the mainland. The Rakata data provided colonization and extinction rates over time, which allowed us to solve for the parameters r_C, r_E, and P. The lineage data presented here do not provide data on the specific rates, and therefore it is more difficult to find estimates for all three parameters. However, the model can still be used to obtain lower and upper boundaries on parameters, which can still be useful! We can consider two extreme cases for the colonization rate, one that is limited by lineage numbers or one that is not constrained by anything but distance (Fig. 2.8). While the potential pool of colonizing species is not known, a minimal estimate might be the pteridophyte flora of the south coast of New South Wales, which has 96 species [38]. In Model 1, we assume that the potential pool is 96, and similar to the Rakata example, we assume that the colonization rate decreases as the number of lineages on the island increases. The other extreme is to consider a maximum potential pool by assuming that the colonization rate is constant as shown in Fig. 2.8, that is, it is independent of the number of species on the island and expressed as R_C. These two proposed models are given by

$$\text{Model 1:} \qquad L_{n+1} = L_n + r_C(96 - L_n) - r_{E_1} L_n, \tag{2.16}$$

$$\text{Model 2:} \qquad L_{n+1} = L_n + R_C - r_{E_2} L_n. \tag{2.17}$$

Since data are given every four million years, we assume that the time step is four million years. Therefore L_1 represents the number of species 4 million years since initial colonization, L_2 represents the number of species 8 million years since colonization, and so on.

Exercise 2.18

What are the units of r_{E_1} and r_C in Model 1 and the units of r_{E_2} and R_C in Model 2? Use the fact that the expression $L_{n+1} - L_n$ is the change in the number of lineages per time step.

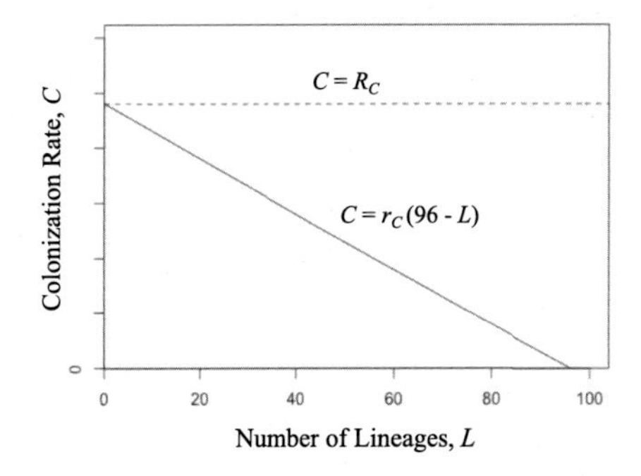

FIGURE 2.8

Assumptions for the colonization rate in Models 1 and 2.

Parameter estimation

Using the lineage data, we can estimate colonization and extinction rates for pteridophytes on New Zealand. However, the approach is slightly different from what we did with the Rakata case. There, we were able to plot the colonization and extinction rates with respect to the number of species and use linear regression to obtain estimates for r_C, r_E, and P. Here, the lineage data do not allow us to do this without specific data on colonization and extinction rates. However, rearranging Eqs. (2.16) and (2.17),

$$\text{Model 1:} \quad L_{n+1} = 96r_C + (1 - r_C - r_{E_1})L_n, \tag{2.18}$$

$$\text{Model 2:} \quad L_{n+1} = R_C + (1 - r_{E_2})L_n, \tag{2.19}$$

it is clear that L_{n+1} is a linear function of L_n in both models, and therefore linear regression can be used to estimate the parameters in each model. In the Rakata problem, note that P was also an unknown, and therefore it would not have been possible to estimate the three parameters r_C, r_E, and P from only two values, slope and intercept. Here, in Model 1 the intercept is $96r_C$, and the slope is $(1 - r_C - r_{E_1})$, and in Model 2, R_C is the intercept, and $(1 - r_{E_2})$ is the slope. Plotting the data points (L_n, L_{n+1}), we can use linear regression to estimate the parameter values in each model!

Exercise 2.19

(a) Using the lineage data, create vectors L_n and L_{n+1} (they must be the same length) and plot L_{n+1} versus L_n.

(b) Perform linear regression on these data and overlay the best fit line on the plot in part (a).

(c) Using the slope and intercept found from linear regression in part (b), estimate r_C and r_{E_1} in Model 1.

(d) Using the slope and intercept found from linear regression in part (b), estimate R_C and r_{E_e} in Model 2.

Model analysis

Now that we have estimates for the model parameters, we can compare the two models, compare the model output to data, and interpret the results.

Exercise 2.20

Setting $L_0 = 1$, use your model parameters and iterate the model (using a `for loop`) to find L_0, $L_1,\ldots,L_9$. Note that Model 1 and Model 2 produce the same output since they both assume linear relationships between L_{n+1} and L_n. Plot the output of the model (L_n versus time), along with the data (number of lineages versus time). How well do the two models capture the observed dynamics?

Exercise 2.21

For each model, use the results of your parameterization to plot the colonization rate as a function of the number of species, that is $C = R_C$ and $C = r_C(P - S)$ versus S. Explain (both mathematically and in the context of this problem) the differences between the assumptions of these rates.

Exercise 2.22

Eqs. (2.18) and (2.19) result in the same model output. However, the different assumptions in the colonization rates result in different values for the extinction rate constants. Using your estimates for the extinction rate constants, compute $1/r_{E_1}$ and $1/r_{E_2}$. What are the units of these expressions? What are your interpretations of these expressions? Why is it important to consider both models?

Exercise 2.23

Write the expression for the equilibrium number of lineages for each model by setting $L_{n+1} = L_n = \hat{L}$ and solving for $\hat{L}$. Use your estimated parameter values to find the equilibrium number predicted by the model. Based on the model prediction, has equilibrium been reached on New Zealand?

Exercise 2.24

Colonization and extinction rates can vary greatly from one taxon to another. For example, reptiles and amphibians in the West Indies showed similar colonization and extinction rates to pteriphodytes in New Zealand, but a similar analysis of birds of the Hawaiian Islands resulted in much higher colonization and extinction rates. Why might colonization rates be higher for birds? Why might extinction rates be higher for birds?

2.3.5 Back to MacArthur and Wilson: effects of distance and area

Now that we have an understanding of how to apply the model, we can consider how other factors, such as area and distance from the mainland, affect the model. According to the MacArthur–Wilson theory, distance from the mainland impacts the rate of colonization, and the inhabitable area impacts the rate of extinction. We can qualitatively understand how these factors affect these rates and in turn how these effects impact the equilibrium.

Exercise 2.25

Consider the model with a constant colonization rate, that is, $S_{n+1} = S_n + R_C - r_E S_n$. (Under what conditions might colonization rate be constant?) A schematic of the extinction rate ($E = r_E S$) and colonization rate ($C = R_C$) as functions of the number of species are shown graphically below. Use this graph to answer the following questions.

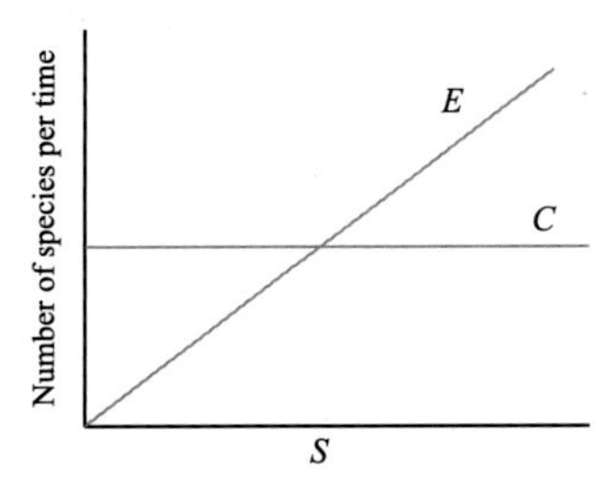

(a) MacArthur and Wilson assumed that the area of an island affects the extinction rate. Consider an island with a large area compared to a smaller one; would the extinction curve for the larger island be steeper or less steep than the extinction curve for the smaller island? Would the equilibrium number of species be different for the larger island? Why? (Revisit Fig. 2.6.)

(b) MacArthur and Wilson also assumed that the distance from the mainland affects the colonization rate. Consider an island closer to the mainland; would the colonization rate curve shift up or down for an island closer to the mainland? Draw it. Would the equilibrium number of species change? How? Why?

(c) We can also represent mathematically the impact on equilibrium by changes in colonization and extinction rates and verify your results from parts (a) and (b). The mathematical model results in an equilibrium given by $\hat{S} = \frac{R_C}{r_E}$. How would changes in area and distance from the mainland affect these rate constants and in turn affect the equilibrium?

Exercise 2.26

Further exploration questions:

(a) To model the impact that area has on extinction rates, we can assume that r_E is a function of island area, but what would this function look like? We can find examples of data that show the relationship between these two parameters (see Fig. 2.9 for an example). In Fig. 2.9, it looks like a decaying exponential function might represent this relationship. Let $r_E = pA^{-m}$, where p is the probability that a species will go extinct on the island, A is the area of the island, and m is a scaling factor for area. Discuss how you would use the data in Fig. 2.9 to obtain an estimate of the scaling factor m.

(b) What other factors might affect extinction rates and how could you use the relationships obtained here to determine if these other factors should be explored?

(c) Many large habitats are being divided into a number of smaller patches of lower total area, isolated from each. This is known as habitat fragmentation. Based on what you have learned, what might be the effect of this on the number of species? What does this suggest for conservation protocols?

(d) What other biological or ecological questions could this model explore?

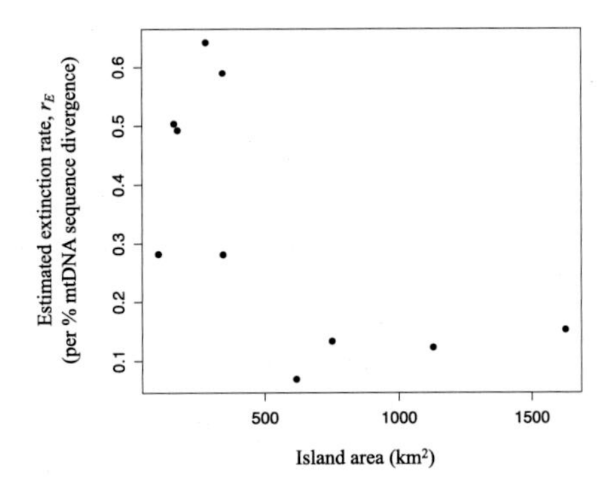

FIGURE 2.9

Estimated extinction rate of nonraptorial land birds on several islands in the Lesser Antilles plotted as a function of island area. Data extracted from Figure 13 in [58].

Our focus has been to explore the simple model set forth by MacArthur and Wilson [40] as an approach to investigate whether an island has reached equilibrium and to understand the observation that different sized islands have different numbers of species at equilibrium. Dispersal properties and niche demands are different among taxa, and these are reflected in the model parameters. Similarly, the importance of other critical features such as niche complexity and species evolution are revealed when data do not fit the model, which leads to a deep understanding of what determines the biodiversity for a particular ecosystem. Refer to Whittaker et al. [73] for a review on advances in integrating classical MacArthur–Wilson dynamics with geoenvironmental dynamics and evolution.

2.4 Case study 2: Pharmacokinetics model

Goals: We will understand the time-course of a single dose of a drug, in this case caffeine, in the body. Using plasma caffeine concentration data, we will use linear regression methods to determine the key parameters, uptake and clearance rate constants, as well as the volume in which the drug distributes. Finally, given these parameters, we will use the calibrated discrete-time model to explore how changes in parameters such as dose, ingestion rate, and body mass alter the kinetics.

2.4.1 Background

Pharmacokinetics: basic concepts and terminology

Pharmacokinetics is the study of how a drug behaves in a body. In other words, it asks and answers the question "How does the body handle a particular drug?" The first step is to describe and understand the concentration versus time profile of a drug in various compartments of the body, for example, the blood or tissue spaces. Once this behavior is known, it is possible to model the therapeutic responses. Here we focus only on the factors that determine the rates and distribution of the drug as it passes through the body. In a typical study, a known amount of drug is administered (the dose), and the concentration (amount/volume) of that drug in the blood plasma is measured at discrete time points. These data are sufficient to estimate kinetic parameters, including the absorption and clearance (elimination) rates and the apparent volume in which the drug distributes.

Pharmacokinetics is highly practical. For example, the pharmacokinetic profiles of a patient's drugs can inform dosage and delivery schedules to ensure a high enough concentration to be therapeutic during the time between doses and far enough from the toxic concentration range to be safe. These studies are the basis for "take two aspirin" but not more often than every four hours and only if an adult. Pharmacokinetics is utilized on a patient specific level, especially when toxic doses are close to therapeutic ones as is the case for most cancer drugs.

Before diving into this case study, we need to become familiar with some fundamental concepts. The kinetic behavior of a drug within the body is commonly divided into four phases: absorption, distribution, metabolism, and excretion. Each of these phases can be described quantitatively. These processes will interact in such a way that there will be a maximum concentration ($C_{\max}$) at a particular time post administration ($t_{\max}$). The former value is critical to know to ensure that toxic levels are never reached.

Many drugs are taken orally as a single dose (amount) in the GI tract followed by absorption or uptake to the blood. When administered orally, both the uptake rate and the fraction of the dose moved from the GI tract to the blood (i.e., its bioavailability) contribute to the overall absorption rate.

The distribution and elimination phases of the model are defined by two parameters, effective volume of distribution V_D and clearance C by metabolic and/or renal mechanisms. The former term V_D refers to the volumes to which it has access, that is, the fluid compartments in which it distributes. Here we will divide the fluid compartments into the blood plasma and extracellular and intracellular fluid (ECF and ICF, respectively) spaces. Any drug that is water soluble will readily exchange between the plasma and ECF. A drug that is sufficiently hydrophobic will readily cross cell membranes and thus also equilibrate with the intracellular spaces and the brain. Thus the volume accessible to the drug will determine its concentration in the plasma, and the latter determines its elimination rate either via the kidneys or by chemical transformation. The relationships among various body aqueous compartments and typical volumes are shown in Fig. 2.10. Use this information to draw a conceptual diagram of the various compartments and the flow of a drug among the compartments.

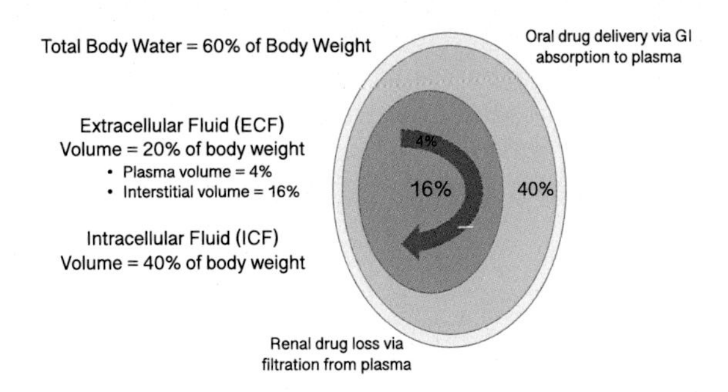

FIGURE 2.10

Distribution of Total Body Water (TBW) among fluid compartments. TBW represents 60% of typical human body weight and is partitioned between intracellular and extracellular spaces. The plasma is the extracellular water contained in the vascular compartment and is in complete equilibrium with the water bathing the cells. Ingested drugs and food molecules enter the plasma from the GI track, distribute through the body depending on their permeability across cell membranes, and exit from the plasma via the kidneys.

How can we determine the effective volume V_D? We know the dose or the amount of drug ingested (denote this by D and assume that the entire dose is absorbed, thus 100% bioavailability), and we can measure the drug concentrations in the plasma. Since concentration is amount per volume and is assumed to be equal in all the aqueous spaces to which it has access when the system is fully mixed, the concentration in the plasma is equal to the concentration throughout the effective volume, which is D/V_D. Some drugs do not cross cell membrane and so never mix with intracellular fluids. It is tricky to get a sample when the drug is thoroughly mixed within the body compartments but is not yet being eliminated. Our model will help us estimate that value. Moreover, for a variety of reasons, the volume may not be exactly equal to one of the known aqueous fluid volumes raising additional kinds of questions to explore when considering how the drug is handled by the body.

Caffeine

One of the most common drugs taken all over the world is something you might not even consider a drug, caffeine. Caffeine has been prized for eons by many cultures as a stimulant that increases alertness and endurance performance. It is a natural plant alkaloid found in coffee, tea, colas, and chocolate as well as an additive to energy drinks and some compound drugs. Like many natural products, caffeine has multiple mechanisms of action. The two most important sources of its effects are due to disruption of signal pathways by blocking adenosine receptors and inhibiting the enzyme phosphodiesterase. In broad strokes, the net outcome of these biochemical effects is first an increase in the release of many neurotransmitters (usually limited by adenosine but unblocked when caffeine blocks its receptors). Secondly, since phosphodiesterase is required to deactivate the "messenger" cAMP generated in response to many signals (e.g., epinephrine on the heart), blocking its action means that any cAMP mediated response will be enhanced (e.g., rapid heart rate).

The chemistry of caffeine is important both for its pharmacokinetics and its mechanisms of action. Caffeine has the dual property of being water soluble and sufficiently hydrophobic to passively permeate the lipid bilayer portion of all cell membranes giving it access to the brain and the ICF spaces of the body tissues. Without this access, it could not reach some of the proteins it affects and would be less useful as a stimulant.

Exercise 2.27

(a) Given this information about the biology and chemistry of caffeine, sketch a graph of the caffeine plasma concentration over time, from the time of ingestion of a cup of coffee until the drug is eliminated. Label when absorption and elimination are occurring. How large do you expect the effective volume to be?

(b) Compare your graph with actual data shown in Fig. 2.11, a typical concentration-time profile generated from 17 healthy human subjects who each drank 100 mg of caffeine in a five minute time period [36]. By examining the figure can you estimate $C_{\max}$ and $t_{\max}$? Which rate seems to be fastest, absorption or elimination? Explain.

(c) Hypothesize what biological factors control the behavior of the plasma caffeine concentration.

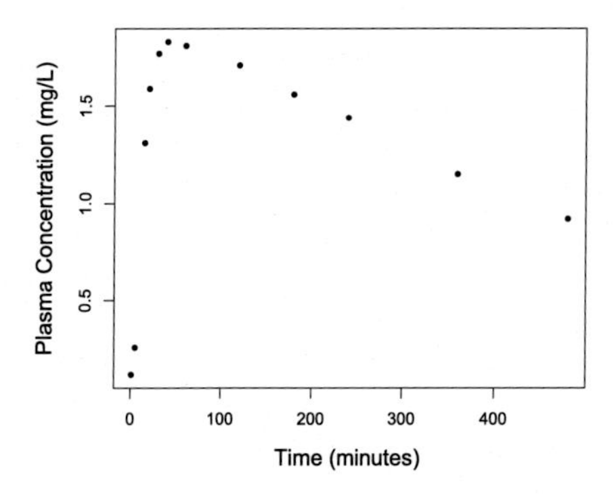

FIGURE 2.11

Plasma concentration after 100 mg caffeine dose taken orally as an energy drink. Data extracted from Fig. 1 in [36].

2.4.2 Formulating the model

Before formulating a mathematical model, it is helpful to develop a qualitative model that depicts the pathway of caffeine through the body. Suppose you consume a caffeinated drink very quickly. The caffeine enters your gastrointestinal tract and then the general circulation from which it can distribute to other parts of the body. Almost 100% of ingested caffeine is absorbed, and therefore we can assume that all of the caffeine consumed will enter the body.

Exercise 2.28

Using the information in the previous paragraph, draw arrows on the diagram below to represent the flow of caffeine. Assume first-order kinetics, that is, the rate out of a compartment is proportional to the amount of drug in that compartment, where k_a (min^{-1}) is the absorption rate constant, and k_e (min^{-1}) is the elimination rate constant. Write the rate constants on the appropriate arrows. Note that the caffeine enters the GI tract almost instantaneously, and therefore there will be no arrow into the GI tract. Instead, we assume that the dose is the initial amount of caffeine in this compartment at time zero. We denote by g_t the amount of caffeine in mg in the GI Tract after t time steps, and denote by b_t the amount of caffeine in mg in the body after t time steps.

GI Tract (g_t)	Body (b_t)

Exercise 2.29

Translate your qualitative model in Exercise 2.28 into difference equations. That is, write g_{t+1} in terms of g_t, and write b_{t+1} in terms of b_t and g_t.

Exercise 2.30

Consider your equation for g_{t+1}. Show that the solution is

$$g_t = g_0(1 - k_a)^t. \tag{2.20}$$

As we have done before (Section 2.1.1), you can show that Eq. (2.20) is a solution by plugging the expression g_t into the left and right sides of the model and showing that they are equal, or you can iterate the model and derive the solution.

Exercise 2.31

In Exercise 2.29, you should have obtained the following equation to represent the amount of drug in the body in terms of absorption and elimination:

$$b_{t+1} = k_a g_t + b_t - k_e b_t. \tag{2.21}$$

We can substitute g_t with the expression in (2.20) to obtain the difference equation

$$b_{t+1} = k_a g_0 (1 - k_a)^t + (1 - k_e) b_t. \tag{2.22}$$

What is your interpretation of g_0? What is b_0?

Eq. (2.22) is a difference equation that models the amount of drug in the body from one time step to the next. As mentioned in the background, many difference equation models are too complicated to obtain explicit solutions. However, Eq. (2.22) is classified as a nonhomogeneous linear difference equation, and therefore we can write the explicit solution, that is, amount of caffeine over time. Eq. (2.22) is linear in the variable b because it can be written in the form $b_{t+1} = h(t) + f(t)b_t$, and it is nonhomogeneous since $h(t) \neq 0$. Why is this important? There are methods to find explicit solutions to these types of equations! In the next section it will be very helpful to have an explicit solution when we need to estimate the parameters. Although the tools to find the explicit solution are beyond the scope of this text, the important part is that we do have the solution. If you would like to learn how to find the explicit solution to Eq. (2.22), refer to [15]. For now, we will give you the solution,

$$b_t = \frac{g_0 k_a}{k_a - k_e} \left[(1 - k_e)^t - (1 - k_a)^t \right]. \tag{2.23}$$

To verify that this is in fact the solution to Eq. (2.22), we can write this solution at time $t + 1$ (the left side of Eq. (2.22)) and check that it does in fact equal the right side of Eq. (2.22).

The solution given in Eq. (2.23) provides an expression for the amount of caffeine in mg in the body at time t. However, measurements of caffeine in the body are in units of concentration, mg/L. Therefore, we need to divide both sides of Eq. (2.23) by the distribution volume V_D in liters. Recall that the volume of distribution (also known as the apparent volume of distribution) accounts for the total fluid volume in which the drug mixes from the plasma (the site of measurement) (Fig. 2.10). Therefore we can rewrite Eq. (2.23) as

$$C_t = \frac{g_0 k_a}{V_D (k_a - k_e)} [(1 - k_e)^t - (1 - k_a)^t], \tag{2.24}$$

where $C_t = b_t / V_D$ is the concentration of the drug in the plasma at time t.

In Eq. (2.24), which parameters will likely be known and which parameters will need to be estimated using a given data set of plasma concentration over time?

2.4.3 Understanding the model

Before finding the parameter values that give the best fit to the data, we provide you with parameter values to investigate the behavior of the model and understand how varying the parameters alters the behavior of the model.

Exercise 2.32

Work through the steps below to understand the output of the model and how changes in the parameter values influence the concentration-time profile.

(a) Assume that 100 mg of caffeine are consumed, $V_D = 45$ L, $k_a = 0.01$ min^{-1}, and $k_e = 0.001$ min^{-1}. Note that V_D, k_a, and k_e are typically estimated using data, but here we provide values to allow you to play with the model. Using Eq. (2.24) and Listing 2.4 as a template, plot concentration of caffeine versus time. (Note that you do not need to use a `for loop` since we have the explicit solution!) Make sure you label your axes and include units.

Listing 2.4 (unit2ex4.R)

```
g0=100; V=45  # g0 is the dose in mg. V is the volume of distribution in liters
time=seq(0,1000) # time in minutes
ka=.01; ke=0.001
C=(INSERT EXPRESSION FOR C_t HERE)
(INSERT COMMAND TO PLOT CONCENTRATION VERSUS TIME)
```

(b) Certain pharmacokinetic parameters are used to characterize the concentration curve. Recall that the peak concentration is denoted by C_{max} and the time at which the peak concentration occurs is t_{max}. Estimate C_{max} and t_{max} from your plot in part (a).

(c) From your plot, what does C_t approach as t gets larger? Verify this by taking the limit of Eq. (2.24) as t goes to infinity. (Hint: What do $(1-k_e)^t$ and $(1-k_a)^t$ approach as t gets larger and larger?) Is this what you would expect biologically?

(d) The elimination rate constant can depend on a variety of factors. Suppose $k_e = 0.005$. How do you expect the graph of C_t to change? Implement this in R and compare the output to your result in part (b), where you assumed $k_e = 0.001$. Explain the difference in terms of how the caffeine is passing through the body.

(e) Now vary the amount of caffeine consumed. What characteristics of the concentration-time profile stay the same and what changes? Can you explain this biologically?

For caffeine, absorption occurs at a much faster rate than elimination. What are the implications of this? The following exercise helps us investigate this question.

Exercise 2.33

(a) Consider the equation for concentration but ignore the absorption dynamic term, that is, let

$$C_t \approx \frac{g_0 k_a}{V_D(k_a - k_e)}(1-k_e)^t \quad \text{for large } t. \tag{2.25}$$

Using the parameter values in Exercise 2.32(a), plot C_t given in Eq. (2.24) and overlay the approximation for large t given in Eq. (2.25). What do you observe?

(b) We can investigate your result in part (a) by thinking about the long-term behavior of C_t (using Eq. (2.24)). Recall that $k_a >> k_e$. Using this fact, explain why C_t can be approximated by Eq. (2.25) when t is large. Explain this in both mathematical and biological terms.

Table 2.4 Plasma concentration after 100 mg caffeine dose administered by oral administration of an energy drink. Data extracted from Fig. 1 in [36].

Time	Plasma concentration (mg/L)
1 min	0.120
5 min	0.259
15 min	1.31
20 min	1.59
30 min	1.77
40 min	1.83
1 hr	1.81
2 hr	1.71
3 hr	1.56
4 hr	1.44
6 hr	1.15
8 hr	0.921

2.4.4 Parameter estimation

Now that we have a better understanding of the model and the model output, we can fit this model to the data given in Fig. 2.11 to see if it captures the observed dynamics. In the study [36], healthy human subjects (9 men, 8 women) were given a maximum of 5 minutes to consume 100 mg of caffeine orally. Blood samples were drawn at the times noted, and the plasma concentrations of caffeine were measured (see Table 2.4). The average age and weight of the subjects were 27.6 years and 75 kg. We can use these measurements of plasma concentration to obtain estimates of the absorption rate constant, elimination rate constant, and volume of distribution. Doing this in two steps, focusing on only the elimination phase and then the absorption phase, will allow us to estimate these three unknown parameters. Recall from Exercise 2.33 that Eq. (2.25) approximates C_t for large t since $k_e \ll k_a < 1$. This means that the long-term behavior is due to only elimination. We can use this approximation to first approximate k_e, and then we will estimate k_a and V_D.

Exercise 2.34

Work through the following steps to estimate k_e.

(a) Recall that we can use linear regression to estimate parameters if the equation is in linear form. Therefore a first step will be to take the log of both sides of Eq. (2.25) to obtain

$$\ln C_t \approx \ln\left(\frac{g_0 k_a}{V_D(k_a - k_e)}\right) + t\ln(1 - k_e) \quad \text{for large } t. \tag{2.26}$$

We have transformed this into a linear equation where the dependent variable is $\ln C_t$, the independent variable is t, the slope is $\ln(1 - k_e)$, and the intercept is $\ln\left(\frac{g_0 k_a}{V_D(k_a - k_e)}\right)$. The

slope and intercept might look complicated, but these are just constants since these expressions include only parameters and no variables!
Since k_a is much larger than k_e, the terminal phase is due to mostly elimination. If we plot the natural log of the plasma concentration data versus time, as observed in Fig. 2.12, then we should see that the terminal phase looks almost linear. Do this! (Only plot the data points. You will add the line later.) Inspect your graph to identify the times at which this is true. Eq. (2.26) should approximate the curve at these times.

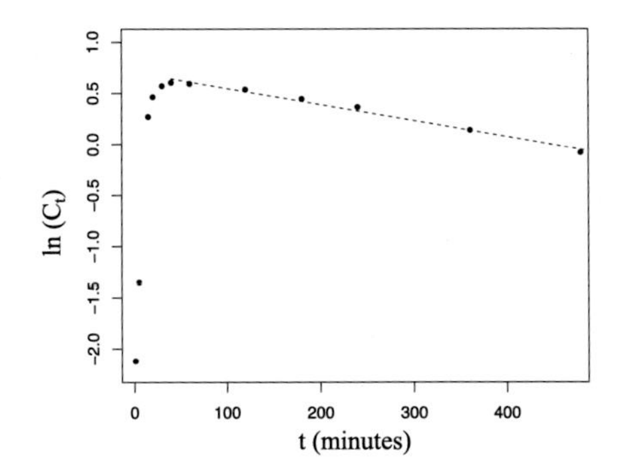

FIGURE 2.12

The natural log of the concentration data (·) is plotted with respect to time. The dashed line shows that the terminal phase can be approximated as a line. This phase is due to elimination only.

(b) Using the template in Listing 2.5, consider the concentration data in Table 2.4 for $t \geq 40$ and take the natural log of this concentration data. Perform linear regression on this data. Why is only a subset of the data (elements 6 through 12) used? Use your results and Eq. (2.26) to determine k_e.

Listing 2.5 (unit2ex5.R)

```
# Data (time in minutes, plasma concentration in mg/L)
t=c(1,5,15,20,30,40,60,120,180,240,360,480)
C_data=c(0.120,0.259,1.31,1.59,1.77,1.83,1.81,1.71,1.56,1.44,1.15,0.921)

# Select data from elimination phase
C_tail=C_data[6:12]
t_tail=t[6:12];

# Linear regression
logC_tail=log(C_tail) # Take ln of data
fit1=lm(FILL IN EXPRESSION HERE)

# Assign parameters from regression results
b=fit1$coefficients[[1]]
slope1=fit1$coefficients[[2]]
ke=WRITE EXPRESSION HERE IN TERMS OF slope1
```

Be sure to note your value of k_e for the next phase of our work. Also, note that linear regression gave you the slope and intercept of Eq. (2.26). We denoted the intercept by b, so that $b = \ln\left(\frac{g_0 k_a}{V_D(k_a - k_e)}\right)$. Using your value of b from linear regression, you now have an approximate value of $\left(\frac{g_0 k_a}{V_D(k_a - k_e)}\right)$, that is, $\left(\frac{g_0 k_a}{V_D(k_a - k_e)}\right) \approx e^b$. Write this value down as well since you will need this soon!

(c) Reproduce Fig. 2.12 using your result from part (b).

Exercise 2.35

Now that we have an estimate for k_e, we can estimate k_a, but it takes a bit more work! Work through the following steps to estimate k_a using the value of k_e found in the previous exercise and the initial data points.

(a) We need to consider the absorption phase, and therefore we need to go back to the original equation for the caffeine concentration,

$$C_t = \frac{g_0 k_a}{V_D(k_a - k_e)}\left[(1 - k_e)^t - (1 - k_a)^t\right].$$

We can do some algebraic manipulations to put it into a form that will allow us to perform linear regression to estimate k_a. Multiply both sides by the factor $V_D(k_a - k_e)/g_0 k_a$ (i.e., $1/e^b$, where b is the intercept from earlier) to obtain

$$\frac{V_D(k_a - k_e)}{g_0 k_a} C_t = (1 - k_e)^t - (1 - k_a)^t.$$

We can rearrange to isolate the exponential term with k_a:

$$(1 - k_e)^t - \frac{V_D(k_a - k_e)}{g_0 k_a} C_t = (1 - k_a)^t.$$

Take the log of both sides to obtain

$$\ln\left((1 - k_e)^t - \frac{V_D(k_a - k_e)}{g_0 k_a} C_t\right) = t \ln(1 - k_a). \tag{2.27}$$

Yikes! This looks messy. However, consider the left side as your new dependent variable. We have values for everything on the left side; we have data points for C_t, an estimate of k_e, and Exercise 2.34 provided us with an estimated value of $\left(\frac{g_0 k_a}{V_D(k_a - k_e)}\right)$. Use the first few concentration measurements in the left side and define these as your output variable and the corresponding time values as your input variable, then apply linear regression to this manipulated set. Use Listing 2.6 to run a linear regression analysis and use these results to find the estimated value of k_a.

Listing 2.6 (unit2ex6.R)

```
# Select data from absorption phase
C_head = C_data[1:5]
t_head = t[1:5]

# Linear regression
log_res = log((1-ke)^t_head-C_head/exp(b)) # Dependent variable
fit_ka = lm(FILL IN EXPRESSION HERE) # Force line to go through origin
slope2 = FILL IN EXPRESSION HERE
ka = WRITE EXPRESSION HERE IN TERMS OF slope2
```

Exercise 2.36

Values for k_e and k_a have been obtained, but the volume of distribution V_D is still unknown. Recall that in Exercise 2.34(c), you found that $\left(\frac{g_0 k_a}{V_D(k_a - k_e)}\right) \approx e^b$, where b was the intercept determined by linear regression. We can rearrange this equation to obtain

$$V_D = \frac{g_0 k_a}{e^b(k_a - k_e)}. \tag{2.28}$$

All the parameters on the right-hand side are now known and can be used to determine V_D, the effective volume for caffeine. The effective volume is dependent on body mass. In fact, V_D is proportional to body mass, that is, $V_D = \alpha \times$(body mass). Using your value of V_D and body mass of 75.7 kg, compute α and include units. Referring back to Fig. 2.10, with what aqueous spaces does caffeine equilibrate? Does this make sense?

2.4.5 Model evaluation/analysis

Now that all model parameters are determined, we can see how well our model fits the data.

Exercise 2.37

Plot the concentration data and overlay the model output using Eq. (2.24). Insert a legend to distinguish the two curves and include labels on the axes. Does the model capture the observed dynamics?

2.4.6 Further exploration

Now that we have an equation to model caffeine pharmacokinetics, we can explore other interesting questions. For example, how do dose amount, rate of consumption, and body mass affect the dynamics? How much would one have to drink to reach a toxic level? How long will the stimulus of a single cup of coffee last and how can that time be shortened? Alternatively, what is the best strategy to stay alert for a long period of time?

Exercise 2.38

Suppose you are watching your neighbor's six-year-old daughter one afternoon, and she asks if she can have the energy drink in the refrigerator. The drink contains 250 mg of caffeine. Why should you adamantly say no (besides the fact that you would have to watch a hyper child!)? Work through the following questions to help you answer this.

(a) Let $k_a = 0.08$ min^{-1}, $k_e = 0.002$ min^{-1}, and $\alpha = 0.6$. Plot the concentration-time profile for body mass values 20 kg, 40 kg, 60 kg, and 80 kg. (Recall, $V_D = \alpha \times$(body mass), where α is assumed to be constant for various values of body mass.) Graph these curves on the same plot.

(b) An average six-year-old weighs 44 lbs, approximately 20 kg. If the six-year-old quickly consumes the entire energy drink, could her caffeine plasma concentration reach a toxic level? Note that toxic levels can vary, but some studies suggest that plasma caffeine concentrations of 15 mg/L or more can lead to nonfatal toxicities. Add a line to your plot in part (a) that denotes the concentration level 15 mg/L (use `abline(15,0)`, where 15 is the intercept, and 0 is the slope) and discuss your conclusions.

Exercise 2.39

We have estimated the peak concentration $C_{\max}$ and the time $t_{\max}$ at which this occurs by looking at the graph of concentration over time. However, we can calculate these exactly and understand what parameters affect these by using calculus. Recall from your calculus course that candidates for maximum or minimum points of a function are found by setting the first derivative to zero.

(a) Consider the concentration function given in Eq. (2.24) as a continuous function. The peak concentration occurs at the maximum, that is, when $\frac{dC}{dt} = 0$:

$$\frac{dC(t)}{dt} = \frac{gk_a}{V_D(k_a - k_e)}\left[(1-k_e)^t \ln(1-k_e) - (1-k_a)^t \ln(1-k_a)\right] = 0, \tag{2.29}$$

where we have used the exponential derivative rule $\frac{d}{dt}a^t = (\ln a)a^t$. Note that Eq. (2.29) has a solution if the expression inside the brackets is zero. Setting this expression to zero and solving for t (which we call $t_{\max}$ here), we obtain

$$t_{\max} = \frac{\ln(\ln(1-k_a)/\ln(1-k_e))}{\ln(1-k_e) - \ln(1-k_a)}. \tag{2.30}$$

This result tells us that altering body mass or altering the amount of caffeine consumed does not change the time at which the peak concentration occurs. Does this seem reasonable? Explain.

(b) The peak concentration occurs at $t_{\max}$, which is described by substituting $t_{\max}$ into Eq. (2.24),

$$C_{\max} = \frac{g_0 k_a}{V_D(k_a - k_e)}\left[(1-k_e)^{t_{\max}} - (1-k_a)^{t_{\max}}\right]. \tag{2.31}$$

How would you expect the peak concentration to change if the amount of caffeine consumed is doubled? How would you expect the peak concentration to change if a person weighs half as much as the original subject? Feel free to try these values and calculate the result.

Exercise 2.40

In the previous questions, we assumed that the caffeine was consumed almost instantaneously. Now suppose the subject is drinking a large coffee that contains 410 mg of caffeine, and he takes a sip approximately every minute. If he drinks his coffee in a span of t_f minutes, then we can assume that he consumes $410/t_f$ mg of caffeine per minute.

Our difference equation model is very conducive to including inputs at discrete times! Here we assume that $g_0 = 0$ and $b_0 = 0$. Going back to our iterative equations that you derived in the beginning, for time steps $t = 0, \ldots, t_f$, the system becomes

$$g_{t+1} = g_t - k_a g_t + \frac{410}{t_f}, \tag{2.32}$$

$$b_{t+1} = b_t + k_a g_t - k_e b_t. \tag{2.33}$$

Note that the equation for the body compartment stayed the same. However, in the GI tract we are adding a dose every minute up to t_f. Once the caffeine is consumed, for $t = t_f, t_{f+1}, \ldots,$ the system is

$$g_{t+1} = g_t - k_a g_t, \tag{2.34}$$

$$b_{t+1} = b_t + k_a g_t - k_e b_t, \tag{2.35}$$

which is our original system. Set up two `for loops` to simulate a 70 kg person drinking 410 milligrams over a span of 20 minutes. Use parameter values $k_a = 0.08$ min^{-1}, $k_e = 0.002$ min^{-1}, and $\alpha = 0.6$. Plot the plasma caffeine concentration over a span of 10 hours. (Recall, you will need to divide b_t by V_D to plot concentration!) Estimate $C_{\max}$ and $t_{\max}$ from your plot.

Exercise 2.41

Similar to the previous exercise, suppose our 70 kg subject sips a large coffee containing 410 mg of caffeine over 20 minutes, starting at 9:00am. However, on this day he also has a double espresso shot containing 150 mg at 11:00am. What is his $C_{\max}$? What is his plasma caffeine concentration at 9pm?

By now, you should have a good sense of how caffeine concentrations vary with time, body size, amount, or coffee drinking strategy. Think about how these kinetics might apply to other common drugs such as ibuprophen (fast vs. slow release) and ethanol.

2.5 Case study 3: Invasive plant species

Goals: We will learn about the complex dynamics of a biennial plant and the features that make it invasive. By understanding its life history we will model how it responds to various management control strategies over a period of years. These results will be used to suggest effective strategies for farmers and conservationists who want to remove garlic mustard from their fields and local forests.

2.5.1 Background

Garlic mustard (*Alliaria petiolata*), now considered a noxious weed, was brought to New York from Europe in the 19th century as a vitamin-rich food source and medicinal plant with antiseptic properties. Since then it has spread across the United States and Canada to the Pacific Northwest and even Alaska, wherever damp forests with alkaline soils occur. One concern is that as an understory plant in deciduous forests, where it can grow to extremely high density, it has the potential to replace native species. It also grows at high densities along railroad lines and in farm fields ruining crop yield. Despite concerted removal efforts, garlic mustard is extremely difficult to eliminate. When a plant persists despite being treated with herbicide or physically removed, there are several possible reasons including persistent roots from which new shoots grow, a constant influx of seeds or a large seed bank. The most likely explanation for garlic mustard persistence is a large bank of seeds in the soil. Our model will let us explore whether this is the case and what it might take to eliminate it.

Garlic mustard has a two-year life cycle. When seeds germinate in early spring, they grow into rosettes (see Fig. 2.13A), some of which will persist through the summer and winter. In the spring the surviving rosettes will mature into adults and flower (see Fig. 2.13B). They are pollinated by insects or self-pollinate forming multiple long narrow seed pods resulting in as many as 600 seeds per plant, depending on local conditions. The seeds disperse by falling within 1–3 meters of the plant. Then the adult dies, but the seeds can survive for up to 10 years. Progression through the life cycle depends on the germination rate of the seeds, the survival rate of the rosettes through the summer and over the winter, and finally on the number of seeds set by the adults. As for many plants, each of these parameters is influenced by the density of rosettes and adults via intraspecific effects. For example, the greater the number of adults (shoots), the lower the survival rate of co-existing rosettes and the fewer number of seeds per adult plant. Similarly, the higher the density of rosettes during the first summer, the fewer seeds each adult will form the next summer, and the fewer rosettes will make it through the fall and winter. The consequences of this competition can be so dramatic such that one year a site will have mostly rosettes (first year plants) and the next year the same area will have primarily adult plants (shoots).

(A)

(B)

FIGURE 2.13

(A) Rosette stage and (B) adult stage of garlic mustard. Photos taken in Asheville, North Carolina.

Exercise 2.42

Consider the life cycle of garlic mustard shown in Fig. 2.14.

(a) On the diagram, indicate the seasons during which the rosettes and adults exist.
(b) List the rates needed to model one life cycle from seed to adult forming seeds.

2.5.2 Model formulation

Garlic mustard has a three-stage life cycle of seeds (S), rosettes (R), and adults (A). Some seeds will germinate into rosettes in the early spring, and the other seeds will remain in the seed bank. Those rosettes that survive from seedlings, then through the summer, and finally through the winter will become flowering adults the following spring. The adults set seed and subsequently die in midsummer. The population count is taken in early summer when rosettes and adults are both present.

This process can be represented by the following difference equations [48,17,49]:

$$\begin{aligned} S_{t+1} &= (1 - g_2)S_t + v(1 - g_1)A_t f, \\ R_{t+1} &= v g_1 s_1 A_t f + g_2 s_1 S_t, \\ A_{t+1} &= R_t s_2 s_3, \end{aligned} \tag{2.36}$$

where v is the first-year seed viability, f is the fertility of adults, g_1 is the proportion of viable seeds that germinate after one winter, g_2 is the proportion of viable seeds that germinate out of the seed bank, s_1 is early survivorship of recently germinated individuals until early summer, s_2 is summer survivorship of rosettes between early

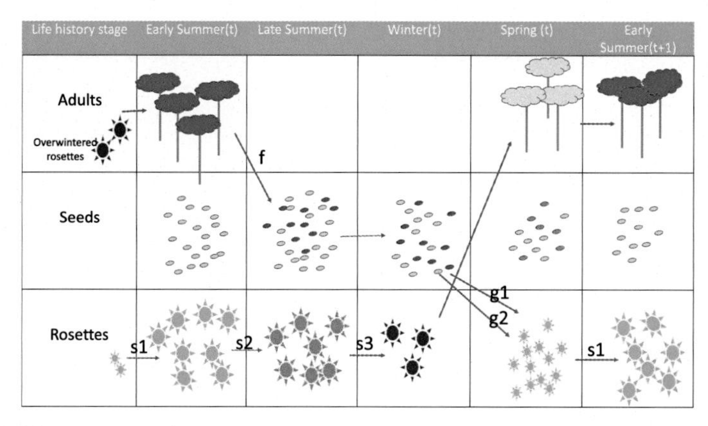

FIGURE 2.14

Life cycle of garlic mustard. The two year life cycle starts with seed germination in the spring to rosettes, that mature and overwinter before developing flower stalks the next spring. These produce seeds and then die back. Germination rates g_1 and g_2 are from new seeds (the summer before) or from seeds in the large seed bank g_2. The number of new seeds is indicated by the fecundity (f). Rosette survival rates are critical to the population dynamics and monitored from spring to early summer (s_1), early to late summer (s_2) and late summer to winter (s_3).

summer and August, and s_3 is winter survivorship of rosettes between August and May. The parameters v, s_1, g_1, and g_2 are based on a set of measurements and appear to be independent of density. However, three of the parameters are not constant in time, but instead they are functions that depend on density. These include fertility of adults (f), summer survivorship of rosettes between early summer and August (s_2), and winter survivorship of rosettes between August and May (s_3).

Exercise 2.43

(a) In your diagram, label the stage at which the census of each population density (S_t, R_t, A_t) is taken.

(b) Which of these parameters are density dependent? Can you speculate as to possible mechanisms by which the density of plants might alter survival or fecundity?

(c) Describe in words what you think a very high population of rosettes one year might do to the number of adults the next year.

The density-dependent parameters can result in oscillatory behavior in which a year of high rosette density is followed by a year of low rosette density and high adult density. In the next section, we will estimate the parameters and iterate the model to compute the number of seeds, rosettes, and adults predicted over time. For now, consider an example of the model output for rosette density and adult density over time in Fig. 2.15, where we observe that each population settles into a steady oscillation.

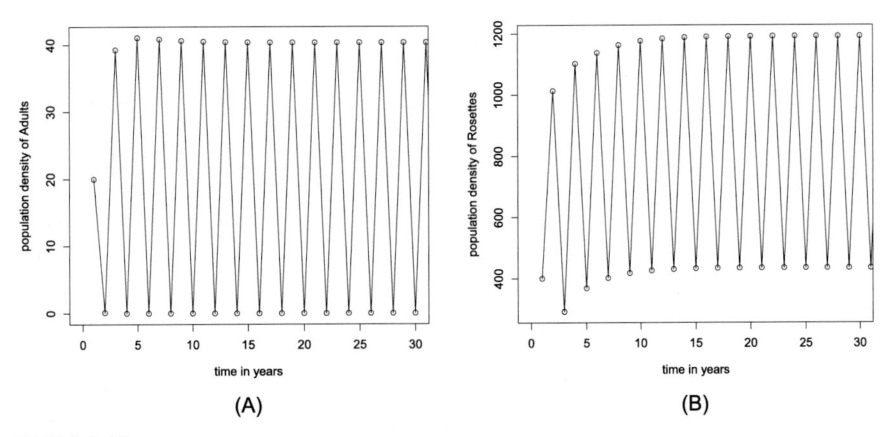

FIGURE 2.15

Examples of model output for rosette density and adult density over time.

Exercise 2.44

(a) Based on the model output given in Fig. 2.15, describe how the populations are behaving from year to year.
(b) Describe how the density dependence of some aspects of the life cycle might generate cyclic behaviors.
(c) Why would it be important to understand the dynamics of garlic mustard when designing a management control plan?

To understand the model further and test control strategies, we must first understand how summer survivorship s_2, winter survivorship s_3, and fecundity f depend on population densities and develop functional forms of these relationships. Luckily, there are data that can help us determine these relationships! In the next section, we will use these data sets to formulate explicit formulas for s_2, s_3, and f and estimate necessary parameter values.

2.5.3 Parameter estimation

Our model contains many parameter values that must be estimated. The values of ν, g_1, g_2, and s_1 were estimated in previous studies [48], and these are given in Table 2.5. However, fertility f, summer rosette survivorship s_2, and winter survivorship s_3 are density dependent and therefore not constant in time. Pardini et al. [48] collected data to understand these dependencies and used functions to represent these relationships. For example, consider their data on fertility shown in Fig. 2.16A. The fertility clearly decreases as the adult density increases. If we plot the natural log of fertility and plot it versus adult density (shown in the inset), then the relationship is roughly linear. This implies an exponential function is a reasonable relationship to represent the data, given by

$$f(A) = \beta_7 e^{\beta_6 A}. \tag{2.37}$$

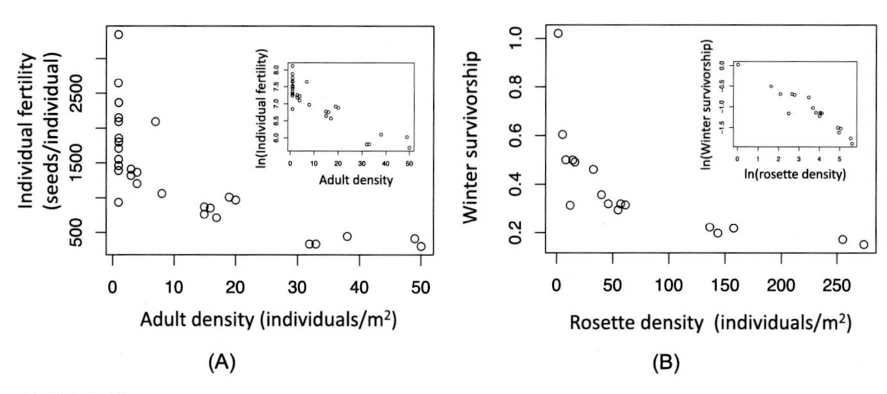

FIGURE 2.16

(A) Individual fertility versus adult density with inset showing the log transformation of fertility versus adult density. (B) Winter survivorship versus rosette density with inset showing the log transformation of each variable. All data were extracted from Fig. 2 in [48].

Exercise 2.45

(a) Given Eq. (2.37) and the data shown in Fig. 2.16A, do you expect β_6 to be positive or negative? Before performing regression, estimate the value of β_7 and provide an explanation for your guess.

(b) Take the natural log of both sides of Eq. (2.37). You should have a linear equation where the independent variable is A and the dependent variable is $\ln f$. What are the expressions for the slope and intercept?

(c) Import the data found in Fertility_ Data.csv. Plot the natural log of the fertility versus the adult density.

(d) Perform linear regression to estimate β_6 and β_7. Then use `abline` to overlay the line of best fit on your plot from part (c). Does this fit suggest that Eq. (2.37) is a reasonable model for the dependence of adult density on fertility?

Now consider the data on winter survivorship versus rosette density (number/m^2) in Fig. 2.16B. It is clear that winter survivorship decreases with increasing rosette density. How do we represent this relationship mathematically? The inset shows the log transformation of each variable, and the relationship is roughly linear. This implies that the raw data can be represented by a power law, such as

$$s_3(R) = (R+1)^{\beta_5}. \tag{2.38}$$

Note that $R+1$ is used instead of R because this forces the curve to be one at $R=0$.

Exercise 2.46

Work through the following exercises to estimate the parameter β_5 in the expression for s_3.

(a) In Eq. (2.38), note that β_5 must be negative to produce a decreasing function as observed in Fig. 2.16B. Given $R \geq 0$, what is the range of possible values of s_3? What does s_3 approach as R goes to infinity? Do your answers seem reasonable based on the physical interpretation of s_3?

(b) Take the natural log of both sides of Eq. (2.38). This is a linear equation with $R+1$ as the independent variable and s_3 as the dependent variable. What is the expression for the slope? What is the expression for the vertical intercept?

(c) Import the data found in Winter_survivorship.csv. Plot the natural log of the winter survivorship versus the natural log of the rosette density plus one.

(d) Perform linear regression to estimate β_5. Then use `abline` to overlay the line of best fit on your plot from part (c). Does this fit suggest that Eq. (2.38) is a reasonable model for the dependence of rosette density on winter survivorship?

We need to use some caution when substituting the expression for winter survivorship (Eq. (2.38)) into the original model. The variable R_t in our model represents the rosette density in the spring. However, the rosette density data in Fig. 2.16B was the density at the end of summer. In terms of our model parameters and variables, the rosette density at the end of the summer is the rosette density at the beginning of the summer multiplied by the fraction that survive the summer, that is, $s_2 R_t$. Therefore, when we substitute our function for s_3 into the model, we will need to use

$$s_3 = (s_2 R_t + 1)^{\beta_5}$$

since the winter survivorship depends on the rosette density at the end of summer ($s_2 R_t$).

The summer rosette survivorship s_2 is more complicated because it depends on both rosettes and adults. The function s_2 can be modeled by the logistic equation

$$s_2(R_t, A_t) = \frac{1}{1 + e^{-(\beta_1 A_t R_t + \beta_2 A_t + \beta_3 R_t + \beta_4)}}.$$

Yikes! We do not have the tools at this point to estimate the parameters in s_2, but they are given in Table 2.5. More details can be found in [48,17].

You can now finish filling out Table 2.5 with your estimates for β_5, β_6, and β_7 from the previous exercises.

2.5.4 Model predictions

Now that all parameter values are estimated, we can iterate the model to investigate the behavior of seeds, rosettes, and adults over time. We will compare two extreme conditions: 1) a seed bank with plenty of seeds but no rosettes or adults in the area and 2) a pristine area with no seeds that has just been colonized with a very small number of rosettes and adults.

Exercise 2.47

(a) In R, define all your parameters from Table 2.5.

(b) Assume that the initial seed density is 1000, the rosette density is 0, and the adult density is 0. Using the template in Listing 2.7, iterate the model and create the following three plots: the seed density versus time, the rosette density versus time, and the adult density versus time. Describe the dynamics that you observe.

Table 2.5 Model parameter values.

Parameter	Value
ν	0.8228
g_1	0.5503
g_2	0.3171
s_1	0.131
β_1	−0.00092
β_2	−0.01612
β_3	−0.00144
β_4	0.1163
β_5	
β_6	
β_7	

Listing 2.7 (unit2ex7.R)

```
t_f = 50 # number of iterations (years)

modS = rep(0,t_f)
modR = rep(0,t_f)
modA = rep(0,t_f)
modS[1] = INSERT NUMBER # initial density of seeds
modR[1] = INSERT NUMBER # initial density of rosettes
modA[1] = INSERT NUMBER # initial density of adults

for(i in 1:(t_f-1)){

  # summer survivorship of rosettes
  s2=1/(1 + exp(-1*(b1*modA[i]*modR[i] + b2*modA[i]+b3*modR[i]+b4)))

  # winter survivorship of rosettes
  s3=(s2*modR[i]+1)^b5;

  # fertility of adults
  f=b7*exp(b6*modA[i])

  modS[i+1]=INSERT EQUATION FOR SEED DENSITY
  modR[i+1]=INSERT EQUATION FOR ROSETTE DENSITY
  modA[i+1]=INSERT EQUATION FOR ADULT DENSITY
  }

PLOT RESULTS
```

Exercise 2.48

Now change the initial condition to zero seeds and a very small density of rosettes and adults. Describe the long-term behavior of seeds, rosettes, and adults. What does this say about the invasive nature of garlic mustard? Do these results confirm the importance of a seed bank as discussed in the introduction?

2.5.5 Management strategies

It should be clear that without implementing a control strategy, once garlic mustard is established, it will persist and have the potential to spread and reach high densities. Moreover, we have shown that its life cycle is complex in ways that yield rebounding populations after a year of lower population density. Thus the model may be a critical first approach for designing effective management strategies that would eliminate or reduce the density of garlic mustard.

Exercise 2.49

Refer back to the life cycle diagram in Fig. 2.14. What methods do you suggest using to reduce the density of garlic mustard, and at what stages of its life cycle could we implement these strategies? What other effects of these methods should be considered (e.g., harming other plant species)?

Given a clear picture of how garlic mustard naturally sustains itself, it is now time to determine if it is possible to successfully implement an induced mortality plan by plant removal strategies that are both feasible and that will limit harm to the plant species that we are trying to protect from being overgrown by garlic mustard. As an example, consider pulling a fraction M_r of rosettes in the spring (second-year rosettes), before they are counted as adults in that year. How would you modify the model in Eq. (2.36) to include this?

Exercise 2.50

The original model can be modified to include the parameter M_r that represents the fraction of rosettes pulled:

$$\begin{aligned} S_{t+1} &= (1 - g_2)S_t + v(1 - g_1)A_t f, \\ R_{t+1} &= v g_1 s_1 A_t f + g_2 s_1 S_t, \\ A_{t+1} &= (1 - M_r)R_t s_2 s_3. \end{aligned} \tag{2.39}$$

(a) Compare the new model to the original model and explain the modification to the model. Also, go back to the original diagram and label where the mortality is taking place.

(b) Modify your code in Listing 2.7 to solve system (2.39). Set $S_0 = 1000$, $R_0 = 0$, and $A_0 = 0$. Try some values of M_r ranging from 0 to 1 and plot the total rosette and adult density $R_t + A_t$ over time. Do you see different dynamics as M_r increases? How large does M_r need to be before effective results seem to occur? What does this say about the survivability of garlic mustard? What does this say about the effectiveness of this control strategy?

(c) We can combine information from the various simulations in part (b) into one graph. Let $M_r = 0$ and run the simulation for 50 years. You will find that the sum of rosettes and adults settles into a steady oscillation between approximately 475 and 1192. We can plot these values as shown below. Pick three more values of M_r, run the simulation, and plot the long-term population by hand on the graph below.

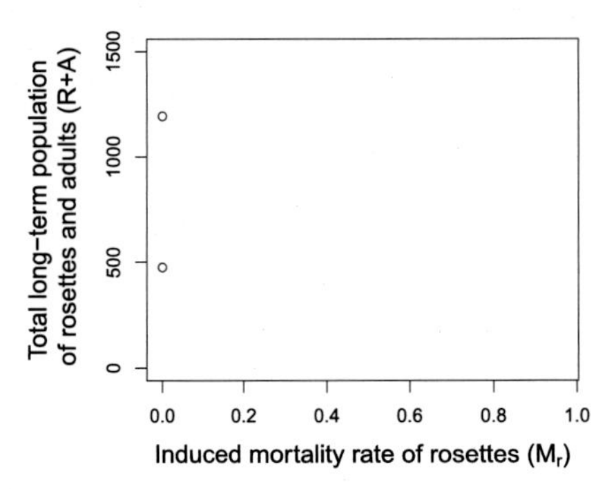

(d) We can automate the process in part (c) using a nested `for loop`! Since multiple values of M_r are used, we can first write this as a vector, and use a `for loop` to repeatedly perform the calculation in part (c). Run the code in Listing 2.8 to produce Fig. 2.17. Make sure you understand the output of each line of code.

Listing 2.8 (unit2ex8.R)

```
t_f = 200 # number of iterations (years)

Mr_vals=seq(0,1,by=0.01) # sequence of Mr values

# Set up empty plot
plot(NULL,xlim = c(0,1), ylim = c(0,1500),xlab = "Induced Mortality Rate of Rosettes",
     ylab = "Long-Term Population of Rossettes + Adults")

# Run the simulation for each value of Mr
for(k in 1:(length(Mr_vals))){

  Mr=Mr_vals[k]
  modS=rep(0,t_f); modR=rep(0,t_f); modA=rep(0,t_f) # Create vectors to store model output
  modS[1] = 1000  # initial density of seeds
  modR[1] = 0  # initial density of rosettes
  modA[1] = 0  # initial density of adults


  # Iterate model far out in time
  for(i in 1:t_f){
    s2=1/(1 + exp(-1*(b1*modA[i]*modR[i] + b2*modA[i]+b3*modR[i]+b4)))
    s3=(s2*modR[i]+1)^b5;
    f=b7*exp(b6*modA[i])

    modS[i+1]=(1-g2)*modS[i] + v*(1-g1)*modA[i]*f
```

```
    modR[i+1]=v*g1*s1*modA[i]*f + g2*s1*modS[i]
    modA[i+1]=(1-Mr)*modR[i]*s2*s3
  }

  # Select populations in the final six years
  FinalR = tail(modR)
  FinalA = tail(modA)
  RnA = FinalR+FinalA

  # Plot long-term values of R+A versus Mr
  points(rep(Mr,6),RnA)

}
```

Here we are only interested in the long-term behavior of the garlic mustard population, and therefore we simulate the model far out in time and plot the population in the last few years (`tail` was used to select the last six values in Listing 2.8). In Fig. 2.17, when $M_0 = 0$, it looks as though two points are plotted instead of six. This implies that the population is oscillating between these two values. For larger values of M_r, it looks as though the populations have approached a steady-state value.

Fig. 2.17 is an example of a *bifurcation diagram*. In mathematics a bifurcation diagram shows the long-term values of the dependent variable as a function of a parameter (called the *bifurcation parameter*) in the system. In Fig. 2.17, we see that for smaller values of the bifurcation parameter M_r, the system oscillates between two values. For larger values of M_r, over time the population will approach a single value. These diagrams are helpful to study the long-term effects of a specific management strategy. For example, based on this model, what fraction of the population has to be removed annually to maintain garlic mustard at a low density? Is it possible to completely remove a garlic mustard population?

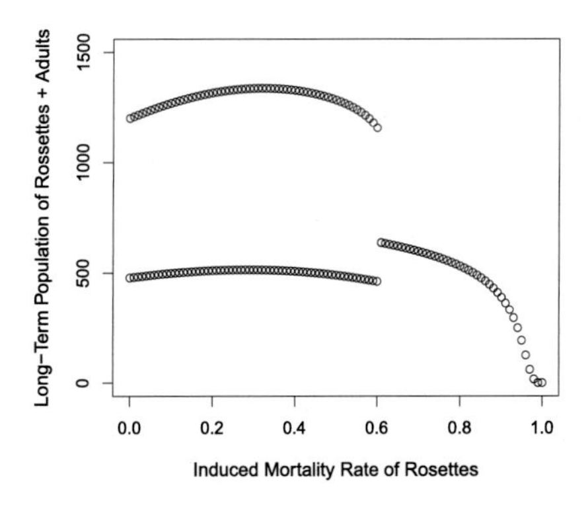

FIGURE 2.17

The effect of pulling second-year rosettes in the spring on long-term garlic mustard rosette and adult populations.

Exercise 2.51

(a) Instead of pulling rosettes in the spring, consider only using herbicide in the fall. Write a modified model to include an annual one-time mortality event (where M_R represents the fraction of rosettes killed by herbicide in the fall) and produce a bifurcation diagram like Fig. 2.17. Discuss your results and which method (pulling all rosettes in the spring or applying herbicide in the fall) is more effective. (Be careful when you modify the model! When rosette mortality is induced in the fall, it will affect only the winter survivorship s_3 and thus the number of rosettes becoming adults in the spring!) Given what you know about summer and winter survivorship estimates s_2 and s_3, do you think it is worth trying to remove rosettes all summer long?

(b) Design your own management strategy, for example, pulling adults or biocontrol agents (e.g., weevils). Compare the effectiveness to the other strategies you explored and discuss the ramifications in implementing each method. Write this up as a report to the local Department of Natural Resources.

2.6 Wet lab: logistic growth model of bacterial population dynamics

Goals: We will learn to measure bacterial population growth in a system of bacteria suspended in nutrient broth over an 8-h period. We will gain an understanding of density-dependent population growth and use linear regression to estimate the intrinsic growth rate and carrying capacity, the key parameters in the logistic growth equation. We will also ask if the initial population size has an effect on intrinsic growth rate or the maximum population size. We may address how other variables, such as nutrient amount or type and temperature or types of bacteria, affect the model parameters in subsequent experiments.

2.6.1 Introduction

Bacteria reproduce by binary fission, in which a single cell divides into two identical cells, each with DNA identical to the parent cell. Bacterial species have their own preferred growth conditions that include the type and concentration of food sources, temperature, and access to oxygen. The *Escherichia coli* that we are studying grow well in a warm nutrient rich environment. Nutrient broth is a commonly used peptide rich medium for growing these bacteria. When the medium is kept at an optimal temperature of 37 °C, *E. coli* may divide every 20 minutes!

Exercise 2.52

Suppose you start with one bacterium/mL. Sketch a graph of population over time that you would expect. What type of mathematical relationship is this?

We will monitor population size by measuring the optical density (OD) or turbidity of the solution using a spectrophotometer. As the number of cells increases, the solution will be more turbid due to light bouncing off of the cells. Therefore, OD (measured as absorbance) is a simple indirect measure of the number of cells. Due to the time frame for bacterial cell division, we will need to sample our populations every 20 minutes for about 6 hours and possibly longer depending on actual growth rates. We will record the OD values and corresponding time, resulting in a data set that is a proxy for cell number/mL (cell density) vs. time. To relate OD to cell number/mL, we will either use a published conversion factor or develop our own relationship by estimating the actual population size using plate counts at several time points.

We will assume a logistic model for bacterial growth and use the data to estimate the model parameters, growth rate, and carrying capacity. Comparing the model output to the data will allow us to evaluate the effectiveness of our model. If the model does not fit the data, then we can adjust our mathematical model using reasonable assumptions. Once we are comfortable with a population growth model, we can alter experimental conditions (e.g., initial density of bacteria, temperature, nutrient concentration) to see what aspects of the growth curve change and how well the model captures the dynamics when one or more of these parameters are changed.

2.6.2 Modeling populations

When a population is provided with unlimited nutrients and space and no threat from predators, it tends to grow at a rate proportional to the population, that is,

$$p_{n+1} = p_n + r p_n, \tag{2.40}$$

where p_n is the population at time step n, and r is the intrinsic growth rate. We can rearrange this equation to

$$\frac{\Delta p_n}{p_n} = r, \tag{2.41}$$

where $\Delta p_n = p_{n+1} - p_n$. On the left side of the equation the numerator is the change in the number of individuals per unit time. By dividing by the number of individuals we obtain the per capita growth rate, which is the rate of growth per individual. In this model the per capita growth rate is constant for any population size. (Recall that a constant per capita growth rate was assumed when modeling the whooping crane population in Section 2.2.)

Exercise 2.53

For $r > 0$ in Eq. (2.41), sketch $\frac{\Delta p_n}{p_n}$ versus p_n. On a separate plot, sketch p_n versus the number of time steps n.

If $r > 0$, then the population increases without bound. When the population is small and resources are abundant, it seems reasonable that the population could grow exponentially. However, as a population gets larger, it is typically constrained by limits of space, nutrients, or accumulation of toxic waste. We would expect the per capita growth rate to decrease as the population increases. The simplest curve that represents this is a line, as shown in Fig. 2.18. When the population is small, the per capita growth rate is close to r, the intrinsic growth rate. As the population increases, the per capita growth rate decreases toward 0 as the population density approaches the carrying capacity K.

Exercise 2.54

Assuming the relationship between per capita growth rate and population is linear (see Fig. 2.18), sketch p_n versus n for the following three cases: initial population p_0 is less than K, p_0 is equal to K, and p_0 is slightly greater than K. (Hint: In the previous example, the per capita growth was a positive constant, and therefore the population was always increasing in time. Here, depending on the initial population, the population can increase, stay constant, or decrease in time!)

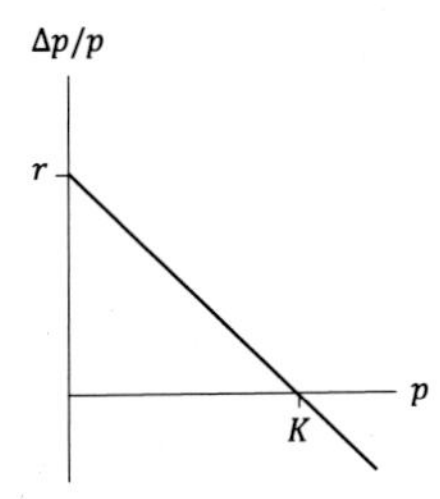

FIGURE 2.18

Model in which per capita growth rate declines as population size increases. The carrying capacity K is defined as the population size at which the per capita growth rate is zero.

Given the relationship in Fig. 2.18, we can translate this into a mathematical equation,

$$\frac{p_{n+1} - p_n}{p_n} = r - \frac{r}{K} p_n, \tag{2.42}$$

where the slope is $-\frac{r}{K}$, and the intercept is r. Rearranging this equation gives the difference equation

$$p_{n+1} = p_n + r p_n \left(1 - \frac{p_n}{K}\right). \tag{2.43}$$

This is called a logistic growth model. Logistic growth curves exhibit an exponential phase when the number of individuals rapidly increases, and a stationary phase when growth is limited by factors such as depletion of an essential nutrient. The logistic model, as with every model, comes with assumptions. Eq. (2.43) assumes the following:

- The per capita growth rate is density-dependent.
- Abiotic variables such as temperature will be held constant in the experiments.
- Reproduction occurs at discrete time steps (bacteria divide by simple binary fission).
- Biotic factors (nutrients, waste) will vary in these experiments, but the mathematical model is agnostic regarding these factors. They are folded together into the carrying capacity.

We will measure the optical densities of a population of *E. coli* over time and use these data to estimate the intrinsic growth rate r and carrying capacity K. By comparing the model output to the data we can determine whether *E. coli* follows logistic growth. Then we will investigate the effect of varying initial population density and/or nutrient concentrations.

2.6.3 The experiment

We will measure population growth of *E. coli* by inoculating a growth medium (nutrient broth) with a small number of cells and observing the change in the number of cells present per mL over time. Each research team will begin with a different initial population density (i.e., cells per mL) so that we can see if the cell growth rate r varies with population density.

We will use one or two methods to measure population over time:

1. Each group will measure the optical density of the bacterial culture using the spectrophotometer at 20 minute intervals over a six hour period giving us our primary growth data.
2. Depending on the preferences of the instructor, you may prepare plate counts early and late in the experiment from serially diluted cultures. This standard method for counting the number of living bacteria will help us correlate OD with the number of bacteria per mL.

Preparation

Exercise 2.55

(a) Make a rough time line of your experiment. Note that the timing and order of work today is critical!
(b) Make a table in your notebook to record optical density.
(c) Draw and label axes prior to lab so that you can plot your data as you go along.
(d) Write down questions you might have before your experiment.

Materials:

- *E. coli* cultures in 50 ml of nutrient broth (in 37°C shaking incubator inoculated with "overnight" culture 1 h before lab)
- Spectrophotometer
- semimicrocuvettes (require 1 mL for accurate reading)
- Timer
- P-1000 and P-200 micropipettes
- Sterile blue and yellow tips
- Tube of sterile broth culture
- Waste container
- Dishpan with dilute chlorox for disposing cuvettes and their bacterial content (one per class).

These items are needed only if doing plate counts.

- 21 sterile microcentrifuge tubes
- Tube racks
- Sterile deionized water
- Plates containing nutrient agar (9 per table)
- Parafilm
- Ice buckets
- Vortex
- Sterile spreaders

Measuring growth rates by optical density

1. About one hour before lab, cultures were initiated; 0.25, 1.0, or 4.0 mL of *overnight culture* were added to 50 mL of broth as labeled. They have been shak-

ing at 37 °C to allow the bacteria to generate the enzymes and transporters they need to grow.

2. Make sure your spectrophotometer is on and set to a wavelength of 600 nm. Use 1 mL of sterile broth in a cuvette (clear faces of the cuvette should be in the light path) as your *blank*. Set the OD of the *blank* to zero.
3. Pick up your culture flask in the shaker and label it with your group name, noting the amount of bacteria added to start the culture.
4. Swirl the culture to mix and pipette 1 mL into another cuvette and insert it into the cuvette holder and close the cover. Read the OD from the display and record it in your notebook as the 0 time reading.
5. Only if you are doing plate counts, take another 0.2 mL from the flask to store on ice in a microcentrifuge tube. Quickly return the flask to the shaking incubator so that it stays warm and well-aerated.
6. Repeat OD reading every 20 minutes over 5–7 hours as needed to get a good data set. By making a rough plot as you go along it is easy to observe when the population is approaching its carrying capacity. OD readings greater than 1.5 are at the limits of the instrumentation.
7. Only if doing plate counts, take two more 0.2 mL aliquots at time points of your choice for plate counting. To get a range of values, try taking one at about 2 hours and after the rate of growth has begun to slow.
8. Clean-up. All cuvettes with bacterial samples can be dumped into a dishpan containing bleach. Flasks and remaining cultures should be collected so that they can be autoclaved.

(Note: This lab can also be done with Baker's yeast. See Technical Notes.)

Measuring growth rate by serial dilution and plate count

This part of the lab is not essential as conversions between OD and bacterial density are available. For these purposes, it is reasonable to assume that an OD reading of 1 represents about 8×10^8 *cells/mL. The value of doing plate counts is to get a strong feel for the meaning of these bacterial density values, understand how OD can be converted to cell density, and learn a technique that might be useful for future projects.*

The bacterial density in our culture flasks ranges from 10^6 to 10^9 cells per mL, more than we can count directly by eye under the microscope. Since a single bacterium will divide to form a visible colony, we can count individuals by adding a diluted sample to a plate and counting the number of resulting colonies 24 hours later. To determine the concentration of living bacteria in our growth flasks, that is, the number of individuals per mL, we will make a set of serial dilutions from which a sample can be plated onto nutrient agar medium that is subsequently incubated for at 37°C for 24 hours so that the cells can divide forming a visible colony to be counted.

We will prepare a set of serial dilutions to decrease the concentration of bacteria to a number we can actually count (e.g., 10–100) when each live bacterium grows into a colony on a plate or petri dish. In the next exercise, we will think about how this works for a series of 10-fold dilutions.

Exercise 2.56

(a) If we take 0.1 mL from a bacterial culture with 2×10^7 cells/mL and add it to 0.9 mL of distilled water or nutrient broth in a sterile tube as shown in Fig. 2.19, then the new density will be 2×10^6 cells/mL. If we mix that thoroughly and then take 0.1 mL from the new tube and add it to 0.9 mL of water, then what is the bacterial cell density in this second dilution?

(b) Remember that we are doing this so that we can grow 10–100 colonies, each representing one bacterium on a single plate. Given that we will use 0.1 mL of diluted sample per plate, how many dilutions will we need if our starting density is 2×10^7 cells/mL?

Serial dilution procedure

1. Label seven of the microcentrifuge tubes (10^{-1}, 10^{-2}, 10^{-3}, etc.) and keep them on a rack to avoid mixing up the tubes.
2. Label the bottom (the side with the agar) of three nutrient agar plates with your group name, date, and the dilution factor. (Write near the edge of the plate because you will count the colonies on these plates.)
3. Add 0.9 mL of sterile nutrient broth or water to each tube. (Make sure not to touch the portion of the pipette tip that will go into the tubes as this may contaminate your culture.)
4. Begin serial dilution by pipetting 0.1 mL (100 μL) of the bacterial culture aliquot and dispensing it into the first tube labeled 10^{-1}. Eject the tip into the waste container.
5. Close the tube and vortex it at medium speed for 2–3 s to mix thoroughly.
6. Pipette 0.1 mL (100 μL) out of this 10^{-1} tube and dispense it into the next tube labeled 10^{-2}, vortex again to mix thoroughly.
7. Continue as indicated in Fig. 2.19.

Plate cultures

Now you are ready to plate your cultures. Because we want to count 10–100 colonies, plate aliquots from the last dilutions, from the tubes labeled 10^{-5}, 10^{-6}, 10^{-7}. Refer to Exercise 2.56 to remember why these dilutions are a good choice.

1. Using new sterile tips, pipette 100 μL aliquot from the 10^{-5} tube, open the plate, and dispense culture onto the surface of the corresponding plate labeled on the bottom as 10^{-5} with your group name.
2. Use a sterile spreader to gently spread the culture all over the agar surface, turning the plate at least three times as you spread to evenly distribute the sample. Cover. Repeat this process for the other two dilutions using new sterile tips and spreaders.
3. When the liquid has soaked in, use parafilm to seal them so that they do not dry out. Place each plate in the incubator at 37 °C with the lid side down to avoid condensation. We will count colonies in a day or two.

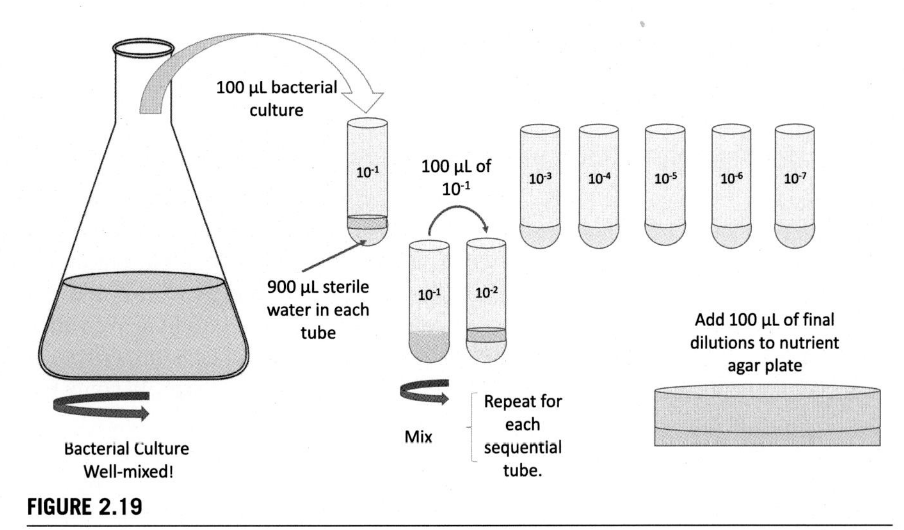

FIGURE 2.19

Diagram of the steps for a series of 10-fold dilutions to reduce the bacterial density sufficiently to be able to count individual colonies on a petri plate.

Plate analysis

To obtain reasonable data, we will count only the plates with approximately 10–100 colonies because this is the range that is considered statistically significant and is actually possible to count accurately.

1. Choose which plates to count, and count and record all the colonies with the dilution factor (e.g., 10^{-6}) in your notebook
2. The number of colony forming units (CFU) per mL in the bacterial culture corresponding to the number of live bacteria per mL can be calculated using the number of colonies, the dilution factor, and the aliquot volume (0.1 mL in our case):

$$CFU/\text{mL} = \frac{(\text{number of colonies})}{(\text{dilution factor}) \times (\text{aliquot volume})}$$

3. Add these values to the class spreadsheet of cells/mL and OD_{600} corresponding to the times at which you took the bacterial samples to be counted.

Conversion between OD and number of cells per mL

We want to relate OD_{600} and cell density so that it is possible to convert your data to cells/mL. We will take the assembled class data and plot cells/mL vs. OD_{600} to determine the relationship, which is likely to be linear at lower cell densities and begin to be sublinear at higher ones. However, the resulting standard curve can be used to estimate actual cell densities, by interpolation, during the entire growth period. A warning: these measurements are technically challenging, and we may have to toss out some values that clearly do not fit the curve. Do not be sad if we cannot use one of your data points.

2.6.4 Model calibration and analysis

We will need to convert the OD_{600} data into cells/mL and then into per capita growth for each time point. First, prepare a spreadsheet with the time points and corresponding OD measurements. Include a description at the top of each column in the first row of the spreadsheet, for example, `Time` and `OD`. Save your file as a ".csv" file. Next to the OD column, make both a `Cells.ml` column and a `PerCap` column. We will use our OD-to-cells conversion to calculate the number of bacterial cells/mL (p) at each time point and then calculate the per capita change in population size. We only need two significant digits.

In your spreadsheet, calculate the number of cells per ml solution (in the `Cells.ml` column) from your OD data column using the formula $p_n = 8 \times 10^8$ cells/mL $\times$ OD_{600} or based on the relationship you determined from your plate counts.

Note that while we denoted the number of individuals by p_n in the introduction, here it represents the density (the number of individuals per ml solution). After you have calculated the density p_n for each time point, calculate the per capita growth rate $\Delta p_n / p_n$ in the `PerCap` column. Recall that Δp_n is the change in population density $(p_{n+1} - p_n)$ in one time step. Each time step is assumed to be 20 minutes; if your data are recorded at varying intervals, then normalize your growth by dividing by the appropriate number of minutes to approximate per capita growth per 20 minutes.

Exercise 2.57

Import the data into R. Plot cell density p_n versus time. Include labels (including units) on your axes and a title in your plot.

To compare our model to the data, we must estimate the values of the parameters r and K. The model is nonlinear, but we can use linear regression if we consider the equation for per capita growth that we developed previously.

Exercise 2.58

The logistic growth model assumes a linear relationship between per capita growth rate and population density. Plot the per capita growth rate data versus p_n as in Fig. 2.18. Include labels (including units) on your axes and a title in your plot.

Exercise 2.59

Perform linear regression on the data plotted in Exercise 2.58. Use the estimated intercept and slope to determine r and K. Overlay the linear regression line on your plot in Exercise 2.58.

Now that we have estimates for the model parameters, we can visualize the raw data over time and our model prediction and compare the results. To obtain the population over time predicted by our model, we need to iterate the recursive equation using our estimated values for r and K.

Exercise 2.60

Using your estimates for r and K from Exercise 2.59, write a `for loop` to iterate $p_{n+1} = p_n + rp_n\left(1 - \frac{p_n}{K}\right)$. Plot the model prediction p_n over time and overlay the raw data on your graph.

Exercise 2.61

Consider your plot from Exercise 2.60. How well does the model fit the data? Does the model exhibit the general dynamics observed in the data? Are there other processes that you think are important to include in the model?

Exercise 2.62

Repeat Exercises 2.57–2.60 using the data from the other two initial population sizes.

Exercise 2.63

Post the values of r and K for each initial bacteria concentration on the class shared spreadsheet. Looking at the class results, is there a relationship between initial bacterial density and r? Is K dependent on the initial bacterial concentration? What might explain these observations? What new questions do they raise?

2.6.5 Experiment part 2: effect of changing media

Another question to explore is how, or whether, growth rate and carry capacity are affected by changes in the bacterial medium concentration. The protocol will be the same as previously, but instead of varying number of starting individuals, we will be varying the bacterial medium concentration from 0.25 to 4 times normal.

Exercise 2.64

What do you predict will happen with more or less concentrated food source? What might be different from your previous results?

For the next set of experiments, it is your job to decide which initial bacterial concentration is the best to use. Varying other conditions may cause r and K to either increase or decrease, so you might want to consider that the accepted values for accurate results of OD lie between 0.1 and 1.5. If your OD values fall outside that range, then they may not be very accurate. Proceed through the experiment as before.

Exercise 2.65

Using your data from Part 2, estimate r and K and plot the model prediction for population over time, along with your raw data. Plot your data points along with the growth curve using your parameter values.

Exercise 2.66

Work through the following questions and provide evidence using your results from the experiments and model analysis.

(a) Does the carrying capacity of each population change or stay constant? Explain.
(b) Does the per capita growth rate change or stay constant as the growing conditions differ? Why?
(c) Did all populations reach their carrying capacities? How do you know? Would you expect these samples to reach a steady state or an equilibrium condition?
(d) Suppose we start with a particular number of bacteria and we would like to maintain that same number of bacteria constant over time. What would we have to do to create that steady-state condition?
(e) Do you think anything else needs to be included in the model to make it more adaptable to various conditions? Explain.

UNIT 3

Differential equations: model formulation, nonlinear regression, and model selection

Learning outcomes

- Formulate differential equation models that represent a range of biological problems and identify assumptions of each model.
- Use sliders to investigate model parameter space.
- Apply nonlinear least squares to estimate model parameters.
- Consider alternative models and use the Akaike information criterion (AIC) for model selection.
- Consider alternative hypotheses and use the tools of parameter estimation and model selection to refine understanding of biological systems.

3.1 Biological background

Biological systems are fundamentally dynamic, dependent on regulated fluxes of materials and energy for survival. Exploring the details of particular fluxes in systems that range from a reaction sequence to an entire ecosystem reveals mechanisms, constraints, and possibilities for that subset of the living world. Thus far, we have used difference equations to model processes at discrete time steps. However, frequently rates of change are not discrete because the variable can take on continuous values (e.g., height), or because the number of elements (individuals, molecules) is large enough, so that the change is effectively smooth and continuous over a defined time period. For example, changes in chemical concentrations due to reactions, fluxes of materials, and changes in sizes of large populations can all be modeled as continuous functions. In this unit, we will use and compare alternative continuous models to study the effect of nitrogen on leaf decomposition rates, to describe the growth of tumors, and to explore parameters controlling predator interactions with prey. The lab of this unit is focused on enzyme-mediated chemical kinetics and on how these may be modulated by toxins or regulators. In each case, we measure changes in amounts or concentrations over time. Once we know the rates, these data can be used to estimate rate constants, affinities, and other limit-

Exploring Mathematical Modeling in Biology Through Case Studies and Experimental Activities
https://doi.org/10.1016/B978-0-12-819595-6.00009-8

ing parameters in the model. Similar to Unit 2, the values of these parameters may be used to interpret the mechanism, predict future events, and make informed hypotheses or control complex systems. Assessing how well the model fits the data and comparing alternative models may also reveal a new understanding of the problem itself and allow us to compare possible mechanisms behind the phenomena described.

After describing a dynamic system with a differential equation model, we can explore the parameter space even before collecting data to assist in fruitful experimental designs. In silico experiments, meaning experiments performed using computer simulations, often follow initial data collection to study the sensitivity of each parameter, that is, the effect of changes in a parameter on the system behavior (e.g., limiting steps and parameters such as the compartment size or uptake rate when considering nutrient delivery) and making predictions into the future. Playing with sliders for parameters and visualizing results is an excellent method for learning about a dynamic biological system and getting a feel for how it will respond to perturbations including its own through feedback mechanisms.

Once we have a sense of how the model behaves, we ultimately want to determine accurate values for each parameter. When modeling biological phenomena, it is frequently necessary to use methods such as nonlinear least squares to attain the best-fit values.

Finally, continuous dynamic models often begin as simple as possible and always have assumptions. Therefore if the model does not fit the data well, then it is likely that some assumptions need revisiting either at the level of the model itself, in the experimental design, or both. Clearly, validating the model or demonstrating that it is a reasonable description of the actual system is critical. Fortunately, it is often feasible to explore alternative models for the same data. Moreover, there are criteria for comparing which model is "best" from a mathematical and biological perspective. Which model is "best" also depends on the application. For example, a simple model of temperature dependence for predation rate may be the best to apply for ecological scale estimates of net prey loss but be completely insufficient to explore temperature sensitive components of each of the steps involved in finding, capturing, ingesting, and digesting a prey item.

We will practice formulating equations to model a number of biological processes at various biological scales, use computational approaches to perform nonlinear regression for parameter estimation, and implement methods to compare how well different models describe a single data set. The experimental lab will introduce the most commonly used biochemical model, the Michaelis–Menten equation to describe enzyme kinetics, and we will collect our own rate data, determine key parameters, and then use the model to explore the mechanism of action of two enzyme inhibitors.

3.2 Mathematical and R background

3.2.1 Differential equation-based model formulation

In Unit 2, we used discrete equations to model changes that occur at natural time steps, for example, the change in a population y may be written in a general form as

$$y_{t+1} - y_t = F(y_t, t, p_1, \ldots, p_m), \tag{3.1}$$

where y_t is the population size at time step t, and the parameters are p_i for $i = 1, \ldots, m$. Rather than denoting the variable as y_t for population size at time step t, we will denote it as $y(t)$ for population size at time t. Now suppose we observe the population at two closely spaced times, t and $t + \Delta t$. If the change in population in one time step is F as seen in Eq. (3.1), then the change in time Δt would be F multiplied by Δt. Therefore, we can write

$$y(t + \Delta t) - y(t) = F(y(t), t, p_1, \ldots, p_m)\Delta t. \tag{3.2}$$

Dividing both sides by Δt, we obtain

$$\frac{y(t + \Delta t) - y(t)}{\Delta t} = F(y, t, p_1, \ldots, p_m).$$

If we take smaller and smaller time steps, that is, let $\Delta t \to 0$, then the left side becomes the definition of a derivative, and we can write it as

$$\frac{dy}{dt} = F(y(t), t, p_1, \ldots, p_m),$$

where $y(t)$ is now a continuous variable. This is an example of a *differential equation*, that is, an equation that relates a function (in this case y) to its derivatives. Note that we can use $\frac{dy}{dt}$ or y' to represent the derivative of y with respect to t.

A continuous variable is appropriate if the state variable is an "amount," for example, concentration. It is also reasonable to use a continuous variable for "number" problems, for example, the number of individuals, if y is sufficiently large such that the addition of one or several is of little consequence and if there are no distinct changes that occur at regular intervals.

In Unit 2, we were introduced to qualitative models (e.g., flow diagrams), which provide a set of instructions for building the formal mathematical model. Sketching relationships is equally valuable when considering continuous processes such as the flux between compartments or conversion from one chemical state to another. Carefully noting units and saying in words what each variable means are a good way to start. However, a qualitative model does not uniquely determine the equations. How does one go from a qualitative model to mathematical equations? In general, we can think about a dynamic model of a variable y as follows:

$$\frac{dy}{dt} = \text{gain rate} - \text{loss rate},$$

that is, the change in a variable is obtained by adding the rate at which y is generated (the gain rate) and subtracting the rate at which y is removed (loss rate). The gain rate and loss rate represent all processes that lead to an increase (synthesis, inflow into compartment, etc.) and decrease (degradation, outflow, etc.), respectively, in y. Sometimes the processes that lead to gain or loss are simple, and sometimes they are complex, but many different biological systems share a few basic processes. Even if two biological systems are on two very different scales, we might use the same mathematical expression to model relationships in both systems!

Here we consider some common biological processes and discuss how to represent them mathematically.

Constant flow rate

The simplest process is a constant flow rate for which the inflow or outflow rate does not depend on time, the relative population, or any intrinsic or extrinsic properties. For example, assuming that species X is produced at a constant rate k, a qualitative diagram representing how the concentration of X, denoted by $x(t)$, changes in time is given by

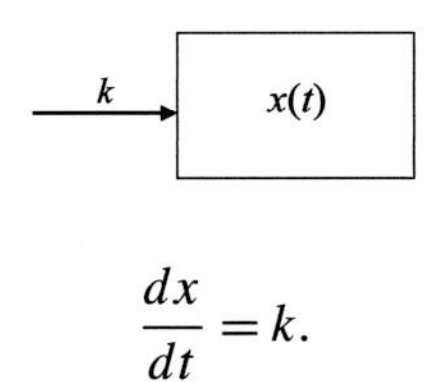

$$\frac{dx}{dt} = k.$$

The rate might be effectively independent of external variables when the source or sink pool is, for all practical purposes, "infinite" and other variables are fixed by the experimenter.

Relative rates

Rates of change typically are not constant in time and instead depend on the current state of the system and/or another variable. For example, the rate at which a protein is inactivated is often dependent on its own abundance. Representing this as a qualitative model, we have

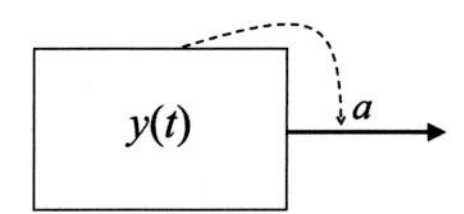

where $y(t)$ is the concentration of active protein, a is the rate constant, and the dashed curve from the box to the arrow shows that the outflow or inactivation rate is dependent on y. Assuming that rate is proportional to the concentration of protein, that is, its concentration decreases linearly with its own concentration, we have

$$\frac{dy}{dt} = -ay,$$

where $a > 0$ is a constant. Since the rate of change is negative when $y > 0$, y is always decreasing. This model is similar to the simple population discrete-time model derived in the beginning of Unit 2, but with only deaths (see Fig. 2.1B). Also, instead of time taking on discrete values, it is now a continuous variable.

Now suppose $y(t)$ represents some population over time, and the population changes due to only births and deaths. If the inflow due to births and outflow due to deaths are assumed to be directly proportional to the number of individuals, then the differential equation model is given by $\frac{dy}{dt} = r_b y - r_d y$. We can rewrite this differential equation as $\frac{dy}{dt} = ry$, where $r = r_b - r_d$, and we can say that the rate of change of y is proportional to y. Dividing both sides of the differential equation by y, we obtain $\frac{1}{y}\frac{dy}{dt} = r$, and we see that r is the *relative* or *per capita* rate. The concept of a relative rate is important in modeling because it often allows us to obtain an experimental measurement of the rate parameter by performing an experiment at a smaller scale than the actual system we are modeling.

Given $\frac{dy}{dt} = ry$, for what values of the parameter r is the population increasing? Decreasing?

Exercise 3.1

Consider a single gene transcribed to mRNA at a constant rate, which then degrades at a rate proportional to the number of copies of the mRNA present. Sketch a qualitative diagram of how the mRNA concentration changes (draw a compartment with inflow and outflow arrows, along with arrows that show what affects these rates), and then write a differential equation that gives the rate of change of the RNA concentration. Define all variables and parameters and state their units.

Exercise 3.2

Suppose we would like to model the rate at which the number of people infected with flu, denoted by $I(t)$, spreads through a community that has a constant total population N. Assume the rate is proportional to the product of the number of people infected and the number of people who are not infected. Write a differential equation that describes the rate of change of the number of infected people. Sketch a graph of how the number of people infected might look like as a function of time.

Mass action

Mass action is a powerful concept that was first derived for chemical systems but is often used in modeling many other types of biological processes. The law of mass action states that the rate of a chemical reaction is proportional to the product of the concentrations of the reactants. For example, using $[X]$ and $[P]$ to denote concentrations of X and P, the rate of product formation and the rate of change of reactant in the reaction

$$X \xrightarrow{k} P$$

are given by

$$\frac{d[P]}{dt} = k[X], \qquad \frac{d[X]}{dt} = -k[X].$$

The derivation of the mass action law relies on statistical mechanics, but intuitively it seems reasonable that doubling the reactant $[X]$ should double the rate of reaction, that is, the amount of product formed per unit time.

If there are two reactants, for example,

$$A + B \xrightarrow{k_1} P_1,$$

then

$$\frac{d[P_1]}{dt} = k_1[A][B], \qquad \frac{d[A]}{dt} = \frac{d[B]}{dt} = -k_1[A][B].$$

What if the two reactants are the same? The rate is still proportional to the product of the reactants. For example,

$$2\,D \xrightarrow{k_2} P_2$$

implies

$$\frac{d[P_2]}{dt} = k_2[D]^2, \qquad \frac{d[D]}{dt} = -k_2[D]^2,$$

where the *stoichiometric* factor 2 appears because each reaction event consumes two molecules of D.

A system often includes a reversible reaction, for example,

$$A + B \underset{k_-}{\overset{k_+}{\rightleftharpoons}} C.$$

Here the rate of change of $[C]$ is equal to the rate of production of C minus the rate of consumption of C, that is,

$$\frac{d[C]}{dt} = k_+[A][B] - k_-[C],$$

and

$$\frac{d[A]}{dt} = \frac{d[B]}{dt} = -k_+[A][B] + k_-[C].$$

Exercise 3.3

What are the units for the two rate constants k_+ and k_-? Explain why they must be different in this example.

Exercise 3.4

Using the law of mass action, write the rate of change of $[ES]$ for the reaction

$$E + S \underset{k_{-1}}{\overset{k_1}{\rightleftharpoons}} ES \xrightarrow{k_2} E + P.$$

Note that this formulation is applicable to a simple enzyme mediated reaction where E represents the enzyme and ES the enzyme-substrate complex. The details of this model will be explored as we prepare for and carry out the lab for this unit.

Although the law of mass action is fundamental in chemical dynamics, it is applied to many other areas, including ecology and epidemiology. For example, the classical Lotka–Volterra predator–prey model is given by the following diagram and corresponding set of differential equations:

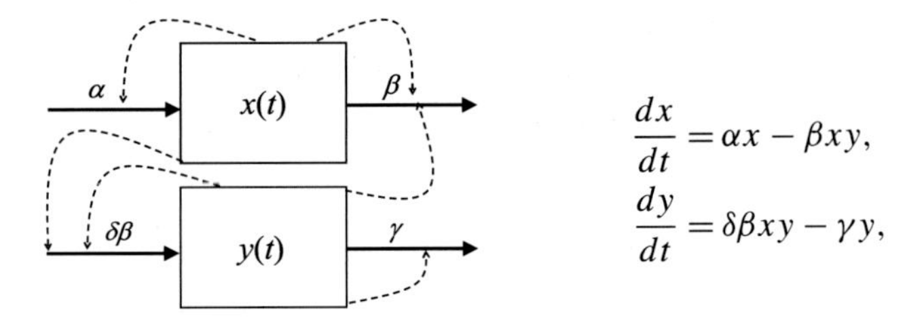

$$\frac{dx}{dt} = \alpha x - \beta xy,$$
$$\frac{dy}{dt} = \delta\beta xy - \gamma y,$$

where x is the number of prey, and y is the number of predators. The prey are assumed to have unlimited resources and will grow without bound if the predators are not present ($\frac{dx}{dt} = \alpha x$ if $y = 0$), and the predators will die out if the prey are not present ($\frac{dy}{dt} = -\gamma y$ if $x = 0$). Now assume that both predators and prey are present, that is, $x, y > 0$. Predation will cause a decrease in the prey population, given by $-\beta xy$, which assumes that the rate of change is proportional to the abundance of prey and abundance of predator. This is the same as mass action kinetics!

Predation should increase the predator population, but by how much? Here the increase is given by $\delta\beta xy$, which is proportional to the mass action term in the equation for the prey. We can understand this term by looking at the units! Note that the units of βxy are (# prey)/time. Since the units of $\delta\beta xy$ must be (# predators)/time, then the units of δ are number of (# predators)/(# prey). Therefore βxy is the number of prey consumed per time, and δ is the conversion of eaten prey into new predator abundance. The factors that contribute to β will be the focus of the predator–prey case study in this unit.

These equations are called the Lotka–Volterra equations, but we can consider a predator–prey model in a more general form, given by

$$\begin{aligned} \frac{dx}{dt} &= \alpha x - f(x)y, \\ \frac{dy}{dt} &= \delta f(x)y - \gamma y, \end{aligned} \tag{3.3}$$

where $f(x)$ is called the *functional response*. Using dimensional analysis, we find that $f(x)$ must have units of prey per unit of time per predator. In other words, it represents the intake rate of predator as a function of the number of prey. In the Lotka–Volterra model, mass action kinetics gives $f(x) = \beta x$, which is referred to as a Type I functional response by population ecologists. In the following exercise, we consider other possible forms of the functional response.

Exercise 3.5

The following questions refer to $f(x)$ in Eq. (3.3).

(a) A Type I functional response is given by $f(x) = \beta x$. Sketch $f(x)$ versus x and discuss the assumptions of this response.

(b) A possible Type II response is $f(x) = \beta \frac{x}{D+x}$. Sketch $f(x)$ versus x and label β and D on your graph. (Hint: What does $f(x)$ approach as x becomes large? What is the functional response when $x = D$?) Then, discuss how this functional response takes into account the fact that predator appetite could become satiated and many predators need time to handle prey (catch, subdue, and consume it).

(c) Now consider an example of a Type III functional response, $f(x) = \beta \frac{x^2}{D^2+x^2}$. Again, sketch $f(x)$ versus x and label β and D on your graph. What types of circumstances might this formulation describe? Discuss the differences (both mathematically and in the context of the predator–prey interactions) between the Type III and Type II functional responses.

While predator–prey models are often used in ecology, these models can be applied to other predator–prey-like systems (e.g., looking at the immune response to tumor cells, the immune cells play the role of the predator, whereas tumor cells play the role of the prey).

Feedback

In a feedback loop a perturbation to a system amplifies the system (positive feedback) or inhibits the system (negative feedback). Positive feedback moves a system away from a steady state. For example, wounded tissue will signal chemicals, which in turn signal platelet activation, and these platelets signal more chemicals that activate more platelets. Negative feedback, on the other hand, stabilizes a system and gives an organism the ability to self-regulate around a set point. For example, when your body temperature rises above its homeostatic level, sweat glands secrete fluid permitting increased evaporative heat loss and blood vessels dilate to allow more of the blood surface area to be exposed to the cooler environment to allow heat loss from the body. If your body temperature goes below its homeostatic level, then shivering generates heat and blood vessels constrict to conserve heat. These mechanisms help the body maintain a homeostatic temperature. Here we provide some examples of positive and negative feedback and how these processes might be expressed as mathematical relationships.

Earlier we considered the simple population model $y' = ry$. If $r > 0$, the rate of change of y increases as y increases, resulting in an unbounded rate of change, which is an example of positive feedback. It is also an example of direct feedback since the variable y influences its own rate.

In Lab 1, we saw that when we started with a small population of bacteria, the rate of change increased because there were unlimited resources. Assuming that $y(t)$ represents the number of bacteria and we now assume that this is a continuous variable, then $y' \approx ry$ when y is small, that is, the population grows exponentially. However, after some time, the population continues to grow, but the limited space and nutrients will cause the rate of change to decrease. Thus the population is inhibiting its own rate of change, which is an example of direct negative feedback. Mathematically, we can represent this as

$$y' = ry\left(1 - \frac{y}{K}\right).$$

When $y \ll K$, $y' \approx ry$, and the population grows approximately exponentially. However, as y increases, the factor $\left(1 - \frac{y}{K}\right)$ plays a greater role. When $y < K$, $y' > 0$, which means that the population is increasing; when $y > K$, $y' < 0$, which means that the population is decreasing. Therefore any nonzero population approaches the carrying capacity K over time, and the model describes how the steady state K is sustained by positive and negative growth fluctuating around the value of the carrying capacity. This is called logistic growth, and this is the continuous version of the discrete logistic growth model you used in lab for Unit 2.

Most biological systems contain a combination of positive and negative feedback loops that interact and that can sometimes produce counterintuitive results! For example, in the predator–prey model in Eq. (3.3), there is direct positive feedback because the rate of change of prey increases as the population of prey increases, but the model also contains indirect negative feedback through the predator–prey interactions. An increase in the number of prey causes an increase in the number of predators, which causes a decrease in the number of prey, which causes a decrease in the number of predators, and so forth! The result is oscillatory behavior for both populations, and negative feedback is essential to produce such behavior. Similar behavior was observed in the garlic mustard case study in Unit 2 due to the density dependence of survival of rosettes.

Many of the processes discussed (constant rates, relative rates, positive and negative feedback, mass action, and saturation) can all be included in a single model! Examples of one such model is given in the following exercise.

Exercise 3.6

Kirschner and Panetta [33] developed a mathematical model of immunotherapy of a tumor:

$$\frac{dE}{dt} = cT - \mu_1 E + \frac{p_1 E I_L}{g_1 + I_L} + s_1, \tag{3.4}$$

$$\frac{dT}{dt} = rT\left(1 - \frac{T}{K}\right) - \frac{aET}{g_2 + T}, \tag{3.5}$$

$$\frac{dI_L}{dt} = \frac{p_2 ET}{g_3 + T} - \mu_2 I_L + s_2, \tag{3.6}$$

where $E(t)$ are the activated immune-system cells (commonly called effector cells) such as cytotoxic T-cells, macrophages, and natural killer cells that are cytotoxic to the tumor cells, $T(t)$ are the tumor cells, $I_L(t)$ is the concentration of IL-2 (a type of cytokine signaling molecule in the im-

Table 3.1 Units and meanings of parameters in Eqs. (3.4)–(3.6) [33].

Parameter	Units	Meaning
c	day^{-1}	antigenicity of tumor
μ_1		decay rate of effector cells ($1/\mu_1$ is lifespan of effector cell)
μ_2		decay rate of IL-2
p_1		maximum rate of effector cell production
p_2	pg·(L·cells·day)$^{-1}$	maximum rate of IL-2 production
g_1		half-saturation constant
g_2		half-saturation constant
g_3	cells	half-saturation constant
s_1		external input of effector cells (treatment term)
s_2		external input of IL-2 (treatment term)
r	day^{-1}	intrinsic growth rate of tumor cells
K		carrying capacity of tumor cells
a		strength of the immune response to tumor cells

mune system) in the single tumor-site compartment, and s_1 and s_2 are treatment terms, in this case the basal rates of effector cell formation and IL-2 formation, which are independent of the model parameters.

(a) You may have very little knowledge about this particular system, but one way of understanding the model is to first identify the variables and parameters. We are given that E and T represent numbers of cells, I_L is a chemical concentration in pg/L, and t is in days. Determine the missing parameter units in Table 3.1.

(b) In Eq. (3.4) the first term on the right side of the equation, cT, represents the production of effector cells where the parameter c represents the antigenicity of the tumor, that is, the ability of the tumor to elicit an immune response. The form cT assumes that the production rate increases without bound as the number of tumor cells increases. Interpret all the other terms on the right-hand sides of Eqs. (3.4)–(3.6).

(c) The mathematical representation of each of these terms you interpreted in part (b) comes along with assumptions of that particular biological process. Discuss these assumptions (e.g., what does the rate depend on? Is the rate assumed to saturate at large concentrations of some variable, or is it unbounded?).

3.2.2 Solutions to ordinary differential equations

Once a differential equation model for some unknown function is formulated, the goal is to solve for the unknown function. In other words, given a differential equation $y' = f(t, y)$, what is the function $y(t)$ that satisfies this equation? For example, the general solution to $y' = ry$ is $y(t) = y_0e^{rt}$, where y_0 is a constant. We can check this by taking the derivative of the proposed solution and verifying that it equals ry. Starting with the proposed solution $y(t) = y_0e^{rt}$, take the derivative of both sides to obtain $y' = ry_0e^{rt}$. The right side ry_0e^{rt} can be replaced with ry since $y = y_0e^{rt}$. Therefore we end up with $y' = ry$, which means that the proposed function is in fact a solution!

In this example, we were given the solution, but how would we obtain solutions to differential equations? In a calculus or differential equations course, we learn methods such as separation of variables to solve particular types of differential equations. However, when modeling biological systems, the models are often complex enough that it is not possible to find an explicit solution. The tumor model in Eqs. (3.4)–(3.6) is an example of such a system. Therefore we must resort to finding approximate solutions using numerical techniques. This will be discussed further in Unit 4. In this unit, we focus on simple differential equation models that do in fact have explicit solutions, but these solutions will be given to us.

3.2.3 Investigating parameter space

One advantage of modeling is the ability to vary parameters and see how these changes affect the model output. Investigating the parameter space can assist in experimental design, help us understand how feedback mechanisms work, and help us predict how the system responds to perturbations in parameters. It is also an incredibly useful tool in determining parameter values that will give us a best-fit model to the data, which will be discussed further in the next section.

If we want to observe how a change in a parameter value affects the output of a model, then we could manually change the parameter value and do this for a number of values. However, the `manipulate` function in RStudio allows us to create sliders to dynamically change values of a parameter. Note that we will need to install this package the first time we use it by running `install.packages("manipulate")`. We only need to do this once, and then every time we open RStudio, we will need to load the library using `library(manipulate)` before using this tool. Look at Listing 3.1 for an example using sliders.

Sliders in RStudio

Consider the exponential function $y(t) = y_0 e^{rt}$. We can create a slider to observe how the shape of the curve changes as we vary values of the initial condition y_0 and per capita growth rate r.

Listing 3.1 (unit3ex1.R)

```
# Load the manipulate library
library(manipulate)

# Specify t values used to evaluate y(t)
t=seq(0,10,length.out=100)

# Use manipulate to create a slider for y0 and r
# Expression to evaluate goes inside braces
# Slider input requires initial and final value of parameter
manipulate({
  plot(t,y0*exp(r*t),type="l",ylab="y",ylim=c(0,1000))},
```

```
y0=slider(0,100,step=1),r=slider(0,.5,step=0.01))
```

Copy this code and run it in RStudio. Move the sliders and observe how the curve changes. Then delete the `ylim` option in the plot command, run the code, and see what happens when you move the sliders. If the sliders ever disappear, then click the gear in the upper left corner of the plot window.

Exercise 3.7

In Exercise 3.5, we studied two functions $f(x) = \beta\frac{x}{D+x}$ and $f(x) = \beta\frac{x^2}{D^2+x^2}$. In this exercise, we will use sliders to study how β and D affect the shapes of these curves.

(a) Set `x=seq(0,50,length.out=100)`. Then, use the manipulate function to plot $f(x) = \beta\frac{x}{D+x}$ and set sliders for $\beta = 0, \ldots, 10$ and $D = 0, \ldots, 15$. What is the effect of changing β? What is the effect of changing D?

(b) Repeat part (a) for the function $f(x) = \beta\frac{x^2}{D^2+x^2}$. In addition, discuss how this curve differs from the curve in part (a).

(c) Now consider the more general form $f(x) = \beta\frac{x^n}{D^n+x^n}$ (referred to as the Hill equation). Use `manipulate` to plot this function and create a slider for $n = 0, 1, \ldots, 10$. Set $\beta = 10$ and $D = 15$. How does the value of n affect the shape of the curve? Why might this function be used to model switch-like responses?

3.2.4 Nonlinear fitting

In Unit 2, you learned how to fit a linear regression model. However, in biology, many relationships are nonlinear. As with linear regression, given data points (t_i, y_i) for $i = 1, \ldots, n$ and a model $y = f(t_i, p_1, \ldots, p_m)$ with parameters p_j for $j = 1, \ldots, m$, nonlinear regression finds the best-fit parameters by minimizing the sum of the squares of the distances of the data points to the model prediction. These distances are called *residuals* and are given by $e_i = y_i - f(t_i, p_1, \ldots, p_m)$, and the sum of these squared residuals is given by

$$RSS(p_1, \ldots, p_m) = \sum_{i=1}^{n} e_i^2 = e_1^2 + e_2^2 + \cdots + e_n^2,$$

where $\sum$ is a convenient notation for denoting the sum of the terms $e_i^2 = (y_i - f(t_i, p_1, \ldots, p_m))^2$ for $i = 1, \ldots, n$. The goal is to find values of the parameters that minimize this sum. Computer algorithms, such as the function `nls` in R can solve nonlinear regression problems by iterating through values of the parameters until the sum of the squared residuals is minimized. Unlike `lm`, when using `nls`, we will need to define the function that we would like to fit, and we will need to include initial guesses for the unknown parameter values. If we have a "bad" guess, `nls` might give an error, or worse, it will give an answer that is not the "best" answer!

Therefore it is dangerous to use built-in functions such as `nls` without understanding what is happening behind the scenes, so we will walk through a simple example to gain a conceptual understanding of nonlinear regression.

Suppose we have four data points of some population over time, denoted by $(t_1, y_1) = (0, 26)$, $(t_2, y_2) = (20, 36)$, $(t_3, y_3) = (40, 78)$, and $(t_4, y_4) = (60, 180)$ as shown in Fig. 3.1A, and we would like to find the best y_0 and r in the function $y(t) = y_0e^{rt}$ that describe these data. We can choose some values for y_0 and r and calculate the sum of the squared residuals, which graphically is equivalent to squaring the vertical distance from each point to the curve and summing these values (see Fig. 3.1A). Choosing different values of y_0 or r would produce different curves, and therefore different sums of squared residuals. Our goal is to find values y_0 and r that result in the minimum sum of the squared residuals.

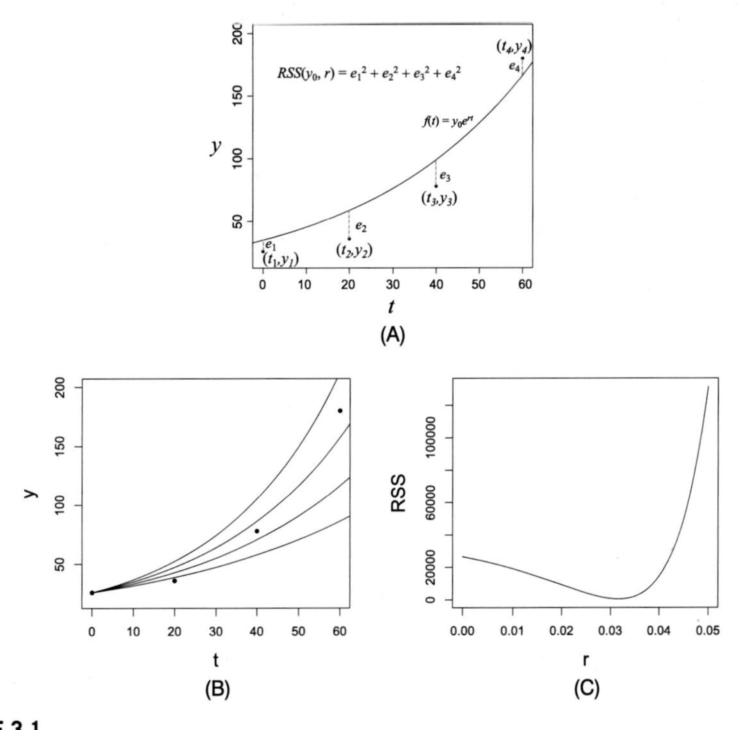

FIGURE 3.1

(A) Data points plotted along with the curve $y(t) = y_0e^{rt}$ for $y_0 = 35$ and $r = 0.026$. Residuals between the curve and data points are shown. (B) Plot data points and the curve $y(t) = y_0e^{rt}$ for $y_0 = 26$ and different values of r. (C) Sum the squared residuals for the given data points and $y(t) = 26e^{rt}$ as r varies from 0 to 0.05.

Suppose we are confident that our measurement error is quite small, and we assume that y_0 is equal to our first data point $y_0 = 26$. This forces our curve to go through the first data point, as shown in Fig. 3.1B. Then our proposed model is $y(t) = 26e^{rt}$, and the only unknown parameter is r. The sum of the squared residuals is given by

$$\begin{aligned} RSS(r) &= e_1^2 + e_2^2 + e_3^2 + e_4^2 \\ &= (y_1 - 26e^{rt_1})^2 + (y_2 - 26e^{rt_2})^2 + (y_3 - 26e^{rt_3})^2 + (y_4 - 26e^{rt_4})^2 \\ &= (26 - 26e^{r\cdot 0})^2 + (36 - 26e^{r\cdot 20})^2 + (78 - 26e^{r\cdot 40})^2 + (180 - 26e^{r\cdot 180})^2 \\ &= (36 - 26e^{20r})^2 + (78 - 26e^{40r})^2 + (180 - 26e^{180r})^2. \end{aligned}$$

Note that the residual sum of squares is a function of the parameter r, which is plotted in Fig. 3.1C. The best estimate of r corresponds to the value when the error between the function and data, that is, $RSS(r)$, is minimized. In Fig. 3.1C, the minimum occurs at approximately 0.03, but how do we find exactly where this minimum occurs? This is a minimization problem like you saw in single-variable calculus. The minimum could be found by finding the derivative of $RSS(r)$ with respect to r and setting it to zero. Alternatively, we could also use the function `nls` in R to numerically calculate the value of r that minimizes the error between the model $y(t)$ and the data. Look at Listing 3.2 for the code to do this.

Nonlinear Regression in R

In the previous example, data points (t_i, y_i) are given, and we are finding the value of r that minimizes the sum of the squared residuals between the data and the model $y(t) = 26e^{rt}$. When using `nls`, the model needs to be specified and an initial guess for the unknown parameter needs to be included.

Listing 3.2 (unit3ex2.R)

```
# Input data and plot data
t=c(0,20,40,60); y=c(26,36,78,180)
plot(t,y,pch=16)

# Use the built-in function nls to find the best-fit parameters
# Include initial guesses for parameters
fit=nls(y~26*exp(r*t), start=c(r=.03))
r=coef(fit)[["r"]]

# Plot the model output
time = seq(0,60,length.out=100)  # Specify a range of input values
lines(time,26*exp(r*time),col="blue") # Overlay model output
legend("topleft", c("data", "model"), col = c(1, 4),lty = c(NA,1), pch = c(16,NA))
```

Input this code into RStudio and run the code. Then change the initial guess in `nls` to `r=1`. What happens?

In Listing 3.2, the algorithm used by `nls` fails if the initial guess (e.g., `r=1`) is far from the actual parameter value. Therefore it is important to obtain a reasonable first guess! How do we determine an initial guess? One method is to calculate RSS for some set of values of r, plot RSS (as shown in Fig. 3.1C) and approximate where the minimum occurs. Another method is to create a slider for r and plot both the model and data, and then move the slider until we have a reasonable fit. Both of these methods are implemented in Listing 3.3 and Listing 3.4.

Obtaining an Initial Guess for One Unknown Parameter

To use nls to find the parameter r that gives the best fit of $y(t) = 26e^{rt}$ to the data, we need to obtain a good initial guess. In Listing 3.3, we evaluate the residual sum of squares for values of r between 0 and 0.05 and plot the residual sum of squares versus r. The output is Fig. 3.1C.

Listing 3.3 (unit3ex3.R)

```
# Data
t=c(0,20,40,60); y=c(26,36,78,180)

r = seq(0,.05,length.out=100) # Specify r values to evaluate RSS
RSS=rep(0,length(r))  # Set up a zero vector to store RSS(r)

# Use a for loop to calculate RSS for each r
for (i in 1:length(r)){
  res = y-26*exp(r[i]*t)  # Vector of residuals for a given r
  RSS[i] = sum(res^{2})  # Sum of squared residuals for a given r
}

# Plot RSS as a function of r
plot(r,RSS,type="l",ylab="RSS")
```

Another method to obtain an initial guess is to plot the data along with the model and use a slider for the unknown parameter. Input the following code into R and play with the slider until it appears that you have the best fit model to the data. You can also adjust the lower and upper values of r along with the step.

Listing 3.4 (unit3ex4.R)

```
time =seq(0,60,length.out=100) # Specify values of t to evaluate model
manipulate({
  plot(time,26*exp(r*time),type="l",ylim=c(0,150)) # Plot model
  points(t,y,pch=20)},     # Plot data
  r=slider(0,.5,step=0.01)
)
```

Each of these methods provides a rough estimate of the unknown parameter, which can then be used as input for nls. The function nls (as shown in Listing 3.2) can then compute the value of the parameter that gives the best fit of the model to the data.

Since the measurement of y_0 may not be accurate due to measurement error, we may want to find estimates of both y_0 and r that give the best fit of the model

$y(t) = y_0 e^{rt}$ to the data. The residual sum of squares is now a function of two parameters, $RSS(r, y_0)$. Again, we could evaluate RSS at a discrete set of values of r and y_0 and plot RSS, which is now a surface in the three-dimensional space (see Fig. 3.2A). However, it is often easier to visualize this as a heat map (see Fig. 3.2B), where the color shows the elevation, that is, the residual sum of squares. The minimum sum of residual squares corresponds to the area that is dark blue, which is still difficult to view due to the large range of values for RSS. To reduce this range of output values, we can plot $\ln(RSS)$ to obtain a map that is clearer in visualizing the location of the minimum (Fig. 3.2C). Using the `plotly` function also allows us to create an interactive map where we can click on that map to reveal the value of RSS at that location. This feature helps us determine the approximate location of the minimum. (Note: We will only use `plotly` in this text to create heat maps. However, for the interested reader, `plotly`'s R graphing library can make a variety of interactive, publication-quality graphs.) An alternative method (similar to the previous example) to approximate values of the parameters is to create a slider for each parameter. Both methods are implemented in R in the following listing.

Obtaining Initial Guesses for Two Unknown Parameters

Similar to Listing 3.3, we first calculate RSS, which is now a function of both y_0 and r. The goal is to estimate the values of both y_0 and r that minimize RSS between the model $y(t) = y_0 e^{rt}$ and the data. We can use a nested `for loop` (see Section 1.2.6) to evaluate the model for combinations of r and y. Run the code in Listing 3.5 in RStudio and hover over the plot with your cursor to find approximate values of y_0 and r that minimize RSS.

Listing 3.5 (unit3ex5.R)

```
t=c(0,20,40,60); y=c(26,36,78,180) #Data

y0 = seq(15,30,length.out=100) # Vector of y0 values
r = seq(0,.05,length.out=100)  # Vector of r values
RSS=matrix(0,length(r),length(y0)) # Matrix to store RSS for combinations of y0 and r

for (j in 1:length(y0)){
  for (i in 1:length(r)){
    res = y-y0[j]*exp(r[i]*t) # Vector of residuals for given values of y0 and r
    RSS[i,j] = sum(res^{2})  # Sum of squared residuals for values y0 and r
  }
}

# Plot ln(RSS)
x = list(title = "initial population (y0)")  # x-axis label
y = list(title = "growth rate constant (r)")  # y-axis label
plot_ly(x=y0,y=r,z=log(RSS),type="heatmap") %>%
layout(title = "ln(Sum of Squared Residuals)", xaxis = x, yaxis = y)
```

Similar to Listing 3.4, sliders are an alternative (and quick!) method to obtain rough estimates of y_0 and r. Run the code below and move the sliders to find approximate values of y_0 and r that give you the best fit.

Listing 3.6 (unit3ex6.R)

```
t=c(0,20,40,60); y=c(26,36,78,180) # Data
time =seq(0,60,length.out=100) # Time vector for model
manipulate({
  plot(time,y0*exp(r*time),type="l",ylim=c(0,200)) # Plot model
  points(t,y,pch=20)},  # Plot data
  y0=slider(0,50,step=1),r=slider(0,.5,step=0.01)
)
```

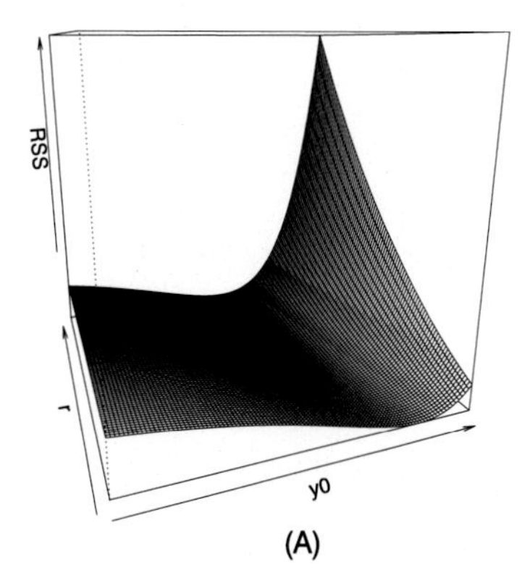

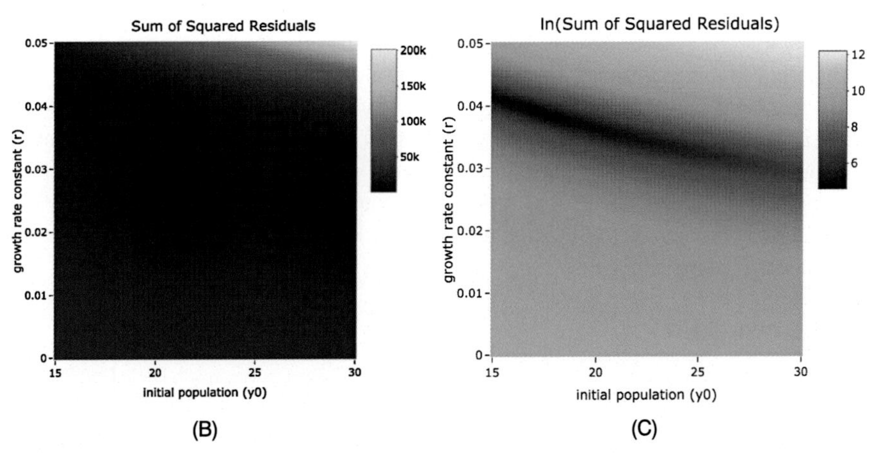

FIGURE 3.2

(A) $RSS(y_0, r)$ surface plotted in 3D. (B) $RSS(y_0, r)$ plotted as a heat map. The horizontal axis is y_0, and the vertical axis is r. (C) $\ln(RSS(y_0, r))$ plotted as a heat map.

Once we have reasonable estimates of y_0 and r, these estimates can be used as input for `nls` as shown in Listing 3.2.

Nonlinear Regression with Two Unknown Parameters

Two parameters can be fit simultaneously in nonlinear regression analysis. Here the formula is included as input to `nls`, where t and y correspond to the data defined previously, and the initial guesses for r and $y0$ are included.

Listing 3.7 (unit3ex7.R)

```
# Nonlinear Regression
fit = nls(y~y0*exp(r*t),start=list(y0=30,r=.03))
fit

# Plot data and model
tMod = seq(0,60,by=.1)  # values of t to use as input into model
yMod = predict(fit,list(t = tMod))  # use results of nls to evaluate model
plot(tMod,yMod,type="l",xlab="t",ylab="y")  # plot model
points(t,y,pch=20)  # plot data
legend("topleft",c("Model","Data"),pch=c(NA,20),lty=c(1,NA))
```

In Listing 3.2, we accessed the parameter r from the output of `fit` and then plotted the function $y(t) = y_0e^{rt}$. Here we used an alternative way to plot the results using the `predict` command.

Why is it important to include good initial guesses for the parameters? Recall that the goal is to find the set of parameters that results in the global minimum of the sum of squared residuals, that is, the smallest error between the model and data. Suppose we are trying to find estimates for p_1 and p_2 in a model $y = f(t, p_1, p_2)$, and the function $RSS(p_1, p_2)$ results in a surface with many local minima (see Fig. 3.3 for an example of a surface with many local minima). If an initial guess is far from the actual global minimum, the optimization algorithm used by `nls` may get stuck in a local minimum (there are many in this function!), and it will give an answer that is not the best one. Therefore it is helpful to use other methods (plotting the residual sum of squares for some sampled set of points or using sliders) to determine approximately where the global minimum occurs. For the interested reader, R has self-starting models that will come up with starting values for parameters to be used in nonlinear regression analysis.

Exercise 3.8

Consider bacteria growth data in Fig. 3.4 (similar to data collected in the growth lab in Unit 2). The data points are

```
t=c(0,1,3,4,5,7,8,12,16,18,20,22,27,28)
y=c(1.05,1.20,1.56,1.80,1.98,2.28,2.61,3.5,4.7,5.10,5.34,5.49,5.58,5.61)
```

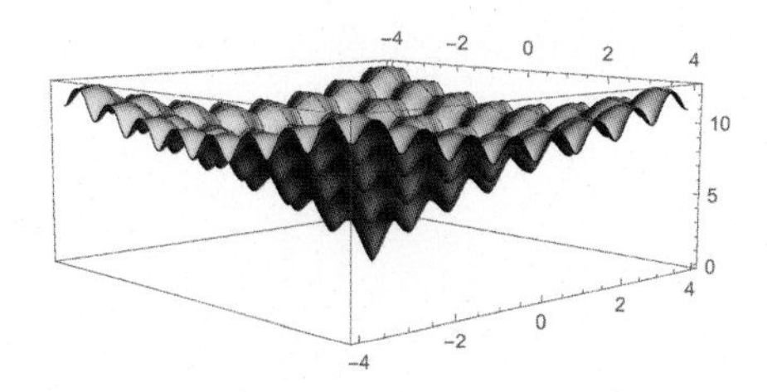

FIGURE 3.3

Example of a surface with many local minima (Ackley function). The Ackley function is widely used for testing optimization algorithms.

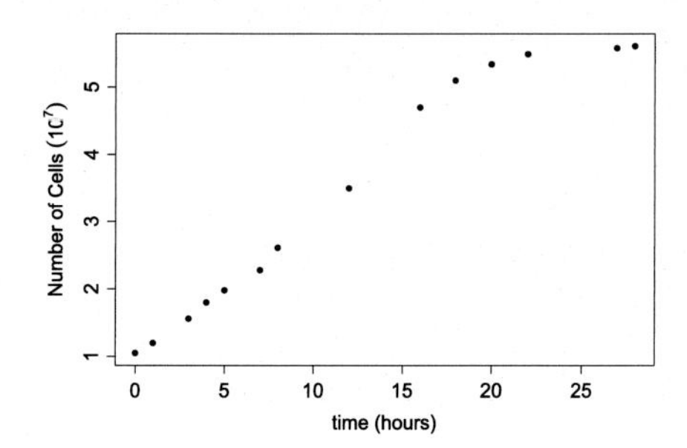

FIGURE 3.4

Time evolution of the number of bacteria cells.

Here, we consider the logistic growth equation $\frac{dy}{dt} = ry\left(1 - \frac{y}{K}\right)$ as a possible model to describe the population growth. The solution to this differential equation model is

$$y(t) = \frac{y_0 K}{y_0 + (K - y_0)e^{-rt}},$$

where y_0 is the initial population size, r is the intrinsic growth rate, and K is the carrying capacity.

(a) The initial number of cells is close to 1, and the number of cells appears to plateau near 5.6, so these can be taken as initial guesses for y_0 and K.[1] To obtain an initial guess for r, use your approximations for y_0 and K to plot both the data and model, and create a slider for r. Report your estimate of r that gives the best visual fit to the data.

(b) Use your estimates of y_0, r, and K as initial guesses in `nls` to find the set of parameter values y_0, r, and K that minimizes the error between the data and the model $y(t)$.

Note that in these examples, nonlinear regression is performed on the analytical solution to the differential equation. However, when using differential equations to model biological phenomena, it is often not possible to find an explicit solution. In

[1] The actual population size is 1×10^8 cells per mL, but we can simply use "1" as long as we remember the actual units when reporting the final data, e.g., the maximum population size is 5.6×10^8 cells/mL. If you do not believe this, then estimate r using both values!

Unit 4, we will learn how to estimate parameters if the solution is not known. For an overview of parameterization, refer to Chapter 4 in Soetaert and Herman [65]. For a thorough discussion on nonlinear regression, see Abraham and Ledolter [1] and Qian [55].

3.3 Model selection

We are often presented with more than one possible model, and we would like to determine the "best" model. Many methods are available for model selection, but here we discuss the Akaike information criterion (AIC). The AIC balances accuracy (how well the model fits the data) with complexity (how many parameters are included). It is possible that including more parameters will provide a better fit to the data, but the increased complexity can decrease the predictive power of the model. Therefore it is important to balance accuracy and complexity. The AIC provides a way to select a model that captures the nature of the outcome variable while avoiding an overfitted model. The AIC is given by

$$\mathrm{AIC} = n \ln(\mathrm{RSS}/n) + 2K, \tag{3.7}$$

where n is the number of data points, and K is the number of parameters in the model. A variant of the AIC is the corrected Akaike information criterion (AICc) given by

$$\mathrm{AICc} = \mathrm{AIC} + \frac{2K(K+1)}{(n-K-1)}, \tag{3.8}$$

which has an additional term to account for small data samples. Note that as the sample size becomes larger, the AICc converges to the AIC.

The actual value of the AIC is not important; the important part is its value relative to the AIC of other models applied to the same data set. A smaller value means that there is a higher level of statistical support for the particular model. Models with AIC differences of less than 2 should be considered equally likely.

Schwarz or Bayesian information criterion (BIC) is another commonly used criteria to evaluate competing models. See Ward [70] for a review of the behavior and assumptions of four commonly used statistical criteria for model selection and refer to Chapter 8 in Haefner [21] for an overview of quantitative validation methods, tests, and indices.

Exercise 3.9

It is well known that aquatic plant photosynthesis depends on temperature, but there is not one accepted model to describe this relationship. Here we will use data on seagrass (*Cymodocea serrulata*) [2], fit two models to the data, and compare the goodness of fit using the AICc. The Yan and Hunt model [77],

$$P(T) = P_{max}\left(\frac{T_{max} - T}{T_{max} - T_{opt}}\right)\left(\frac{T}{T_{opt}}\right)^{\frac{T_{max}-T}{T_{max}-T_{opt}}}$$

has three parameters including the maximum photosynthesis rate P_{max}, the thermal optima T_{opt}, and maxima T_{max} for seagrass photosynthesis. The O'Neill model [64],

$$P(T) = P_{max}\left(\frac{T_{max} - T}{T_{max} - T_{opt}}\right)^{a} \exp\left[a\left(1 - \frac{T_{max} - T}{T_{max} - T_{opt}}\right)\right],$$

has four parameters, where $a > 0$ has no units and no physical interpretation. Using nonlinear regression, we can find estimates of the parameters and then calculate AICc (given in Eq. (3.8)) for each model. Use the template below to do these steps, and discuss your results. Note that the residuals are included in the output of nonlinear regression, which we can then use to calculate RSS!

Listing 3.8 (unit3ex8.R)

```
# Data
P=c(0.78,1.57,2.13,3.05,3.86,3.75,1.08) # Photosynthesis rate
T=c(16.96,21.92,25.92,29.99,33.91,37.92,42.95) # Temperature

#Yan and Hunt Model
fit_yh=nls(P~Pmax*(Tmax-T)/(Tmax-Topt)*(T/Topt)^((Tmax-T)/(Tmax-Topt)),
           start=list(Pmax=4,Topt=35,Tmax=50))
RSS=sum(resid(fit_yh)^{2})
n=FILL IN VALUE; k=FILL IN VALUE;
AICc_yh=FILL IN EXPRESSION

# O'Neill Model
fit_o=nls(P~Pmax*((Tmax-T)/(Tmax-Topt))^a*exp(a*(1-(Tmax-T)/(Tmax-Topt))),
          start=list(Pmax=4,Topt=35,Tmax=45,a=1))
RSS=FILL IN EXPRESSION
n=FILL IN VALUE; k=FILL IN VALUE;
AICc_yh=FILL IN EXPRESSION
```

3.4 Case study 1: How leaf decomposition rates vary with anthropogenic nitrogen deposition

Goals: Leaf decomposition data from three tree species are given along with two proposed models. For each species, there are control and nitrogen treated data sets. We will calibrate these models using nonlinear regression and compare them using the adjusted Akaike information criterion. Using one of the models, we will compare the rate and extent of decomposition between the low nitrogen (control) and high nitrogen treatments for the three tree species.

3.4.1 Background

Decomposition of biological material is an important means of nutrient cycling. When biological materials such as dead leaves fall to the ground, they begin to decompose. This means that their large organic compounds such as proteins, cellulose, and lignin are physically and chemically broken into smaller inorganic and organic molecules and CO_2, generally by an assemblage of microorganisms in the soil. The smaller compounds are then available as nutrients for other organisms. Some material is recalcitrant and breaks down extremely slowly so that these large carbon-containing compounds tend to accumulate in the soil. Decomposition is facilitated by microbial biochemical activity, and the rate and extent of decomposition are dependent on the material and environmental factors such as soil temperature, moisture, pH, and the presence of reactive nitrogen compounds. Slow rates and low extents of decomposition tend to leave more carbon trapped in the soil, whereas high rates and extents contribute CO_2 to the atmosphere. Thus estimates of these rates are one critical component of all climate change models.

Atmospheric reactive nitrogen (such as nitrate and ammonium) levels have more than doubled due to fossil fuels, man-made fertilizers, and sewage from human and animal sources [19,16]. These compounds have spread globally, deposited by precipitation in oceans and landmasses, and we are just now measuring their effects. Changes in the total nitrogen compound concentrations in soils in north temperate regions have led to studies on the effects of increased proportions of atmospheric reactive nitrogen compounds on organic compound deposition.

One study showed the surprising result that simulated increases in anthropogenic nitrogen deposition in forests resulted in greater soil carbon compound retention. Increases in soil carbon compound retention are caused by one of two factors, increases in leaf biomass deposition or decreases in leaf decomposition. It is interesting that leaf litter deposition rates did not increase, suggesting that something about leaf decomposition had changed to cause more carbon to remain in the soil. Two possible explanations exist: either slower leaf decomposition rates or a reduced extent of decomposition (in other words, nitrogen could lead to a slower overall rate of decomposition, or decomposition could occur at the same overall speed, but nitrogen causes decomposition to stop early). Whittinghill et al. [74] investigated this question.

In this case study, you will use data from leaf litter decomposition experiments under low nitrogen and high nitrogen compound deposition. We present two possible models to describe leaf litter decomposition. You will calibrate both models using nonlinear regression and discriminate between them using the adjusted Akaike information criterion (AICc). Then you will investigate the effect of increased levels of nitrogen on the rate versus extent of decomposition.

3.4.2 The data

The data are taken from leaf litter decomposition experiments performed in ambient and elevated nitrate nitrogen deposition environments to mimic the effect of anthropogenic nitrate addition to the soil. Tree species include *Betula alleghaniensis* (yellow birch), *Pinus resinosa* (red pine), and *Acer rubrum* (red maple). Data are given in Tables 3.2a–3.2c.

Exercise 3.10

Plot the data from both treatments. Make a separate plot for each tree species and write down your observations. Discuss any differences you observe between the low and high nitrogen cases and among different tree species.

3.4.3 Model formulation

Before creating a mathematical model, it is helpful to think conceptually about the process of interest, that is, the degradation of large organic molecules to smaller soluble organic compounds and finally to fully reduced (CH_4) or fully oxidized (CO_2) gases, which are released into the atmosphere. Degradation decreases the amount of organic material in the soil, that is, the variable measured here. Mixed microbial species carry out the decomposition, and their ability to do so depends on external conditions, in this case, the extra nitrates added to the soil. Since there is a mix of plant compounds, the rates of decomposition may vary, and there may even be some carbon compounds that are never degraded.

Here we propose two differential equation models, a single exponential (Model 1) and an asymptotic model (Model 2):

$$\text{Model 1:} \quad \frac{dm}{dt} = -r_1 m, \qquad m(0) = 100,$$

$$\text{Model 2:} \quad \frac{dm}{dt} = -r_2(m - M), \qquad m(0) = 100,$$

where m is the percentage of initial dissolved organic carbon (DOC) remaining, t is the sampling time in months, and M is the dissolved organic carbon that is recalcitrant, that is, not susceptible to degradation. These two differential equations are simple linear models, and we are able to find explicit solutions given by

$$\text{Model 1:} \quad m(t) = 100e^{-r_1 t},$$

$$\text{Model 2:} \quad m(t) = (100 - M)e^{-r_2 t} + M.$$

Table 3.2 Data from leaf litter decomposition experiments [74].

Time (months)	% Mass Remaining (low N)	% Mass Remaining (high N)
0	100	100
9	76	75
12	56	65
21	53	59
24	47	54
33	39	48
36	39	47
45	31	50
48	32	36
57	27	48
60	22	34
69	29	37
72	19	35

(a) Yellow Birch

Time (months)	% Mass Remaining (low N)	% Mass Remaining (high N)
0	100	100
9	84	84
12	73	74
21	57	62
24	51	59
33	54	56
36	41	50
45	32	38
48	30	36
57	33	44
60	28	33
69	27	34
72	26	33

(b) Red Pine

Time (months)	% Mass Remaining (low N)	% Mass Remaining (high N)
0	100	100
9	65	67
12	51	58
21	45	53
24	45	54
33	45	49
36	52	55
45	44	53
48	43	54
57	41	53
60	41	56
69	51	56
72	40	48

(c) Red Maple

Exercise 3.11

Determine the units of the parameters r_1, r_2, and M and describe the physical interpretation of each parameter. Also, discuss the assumptions of each model. It might be helpful to sketch $\frac{dm}{dt}$ versus m and/or $m(t)$ versus t for the two models.

3.4.4 Parameter estimation

We are not able to directly obtain measurements of the parameter values. However, we can use the data collected to calibrate the model by using nonlinear regression to find the best estimates for these values. Work through Exercises 3.12–3.13 to obtain estimates for r_1, r_2, and M for the Yellow Birch leaves.

Exercise 3.12

Input the yellow birch low nitrogen data in Table 3.2a or import the file yellowbirch.csv. In part (c), we will use the data to perform nonlinear regression on Model 1. However, recall that `nls` requires a good initial guess for the parameter. First, we will implement two different ways to obtain an estimate of r_1 in parts (a) and (b), and then we will use this estimate as input for `nls`, which will give us the best-fit value of r_1.

(a) Since Model 1 has one unknown parameter, there is a quick and easy way to obtain a rough estimate for the parameter r_1. Choose any data point (t, m), plug it into the model $m = 100e^{-r_1 t}$, and solve for r_1.

(b) We can also use the manipulate function as an alternate method to obtain an estimate for r_1. (Refer to Listing 3.4 for an example using `manipulate`.) Within the `manipulate` command, plot the raw data and the model $m(t) = 100e^{-r_1 t}$ on the same graph and create a slider for r_1. Move the slider until you have a reasonable fit. Report this value of r_1. Is it close to what you found in part (a)?

(c) Now that we have a reasonable estimate of r_1, we can use this as input into `nls` to find the value that results in the best fit (minimizes the sum of the squared residuals) between the data and model. (Refer to Listing 3.2 for an example using `nls`.) Use your estimate of r_1 from part (a) or (b) as an initial guess in `nls`. What is the best-fit value for r_1? How does this compare to your estimates in parts (a) and (b)?

(d) Using the best-fit value for r_1, plot the data and model output on the same plot. Make sure you label the axes and include a legend.

Exercise 3.13

Use the low nitrogen data from yellowbirch.csv and follow the steps below to find the best-fit parameters r_2 and M for Model 2.

(a) Use the manipulate function to create sliders for r_2 and M. Move the sliders until you have a reasonable fit. Report your values of r_2 and M.

(b) Use your estimates in part (a) as initial guesses in `nls`. What are the best-fit values for r_2 and M? How do these compare to your estimates in part (a)?

(c) Using the best-fit values for r_2 and M, plot the data, and model output on the same plot.

(d) Compare your plot from part (c) to the plot in Exercise 3.12(d). Just by looking at it, can you decide if one model is better than the other? Why or why not? Be explicit.

3.4.5 Model evaluation

Now that the best-fit parameters have been calculated for the Yellow Birch data set, is Model 1 or Model 2 better? As discussed in the background, there are various quantitative ways to discriminate between the applicability of Model 1 and Model 2. Here we will use the adjusted Akaike information criterion (AICc).

Exercise 3.14

Using the low nitrogen Yellow Birch data set, calculate and record the AICc (refer to Eq. (3.8)) for Model 1 and Model 2. What conclusions can you make?

Exercise 3.15

Now find the best-fit parameters and calculate the AICc for Models 1 and 2 for the Yellow Birch high nitrogen data, and also for both the low and high nitrogen cases for the remaining types of trees, Red Pine and Red Maple (data in redpine.csv and redmaple.csv). You can use the same code that you used for the previous exercises! You will only need to change the data set you are importing. Report your values for all three tree species. Make a table to report your results.

Exercise 3.16

Compare the AICc for each set of data. Which model is better? Why?

Exercise 3.17

Compare the best-fit parameter values between the low nitrogen and high nitrogen for each type of tree estimated from your preferred model. What conclusions can you make? Discuss your results in the context of the original objective of investigating the effect of increased levels of nitrogen on decomposition rate versus extent of decomposition.

3.5 Case study 2: Exploring models to describe tumor growth rates

Goals: Four different mathematical models are proposed to describe the growth of tumors derived from lung tissue. We will discuss why each of the growth models might be suited to represent solid tumor growth, and then we will calibrate each model to the data using nonlinear regression and evaluate their relative descriptive power using the AICc. Finally we will use a subset of the data to test the predictive power of each model.

3.5.1 Background

Tumor growth patterns have the potential to provide extremely valuable clinical and preclinical information. If growth rates can be easily monitored and effectively modeled, the trajectory of a patient's disease status, the effectiveness of a drug regime or even the history of the patient's tumors might be determined. As tumors begin from a single rapidly dividing cell, the first thought is that growth will be exponential. However, the cellular environments within solid tumors change as they expand with the interior cells having fewer nutrients, often less oxygen and more contact with like-neighbors. The microenvironment is dynamic, and what is sensed by the cells on the tumor periphery may also change in response to the tumor. Immune cell recruitment should inhibit growth rate, whereas other cues may spur tumor growth. We already can see that it is likely that tumor growth rates will not be constant as the tumor enlarges.

To begin to understand a relatively simple tumor growth pattern, Benzekry and colleagues grew and measured lung tumors in mice [4] (data in Table 3.3). These data may be explored using the models listed in Table 3.4 to assess the importance of various factors that might confound the growth rate. Two of the models are very familiar (exponential and logistic growth), and two have been chosen for consideration because they reflect some known aspects of tumor growth. Since long-term projection of tumor growth also is important for patient assessment, we explore how well only a subset of the data, just the early points, will calibrate each model in comparison with the full data set to decide which of our models might be best suited for projecting future growth.

3.5.2 The data

Lung tumors were grown under controlled conditions in mice by injecting a fixed number of murine lung cancer cells under mouse skin [4]. From this location the tumors could be accessed with calipers to measure the smallest and largest tumor diameters, from which they estimated the volume assuming an ellipsoid (see Fig. 3.5). The data in Table 3.3 represent the average volumes from twenty separate lung tumor growth experiments, each followed for 21 days.

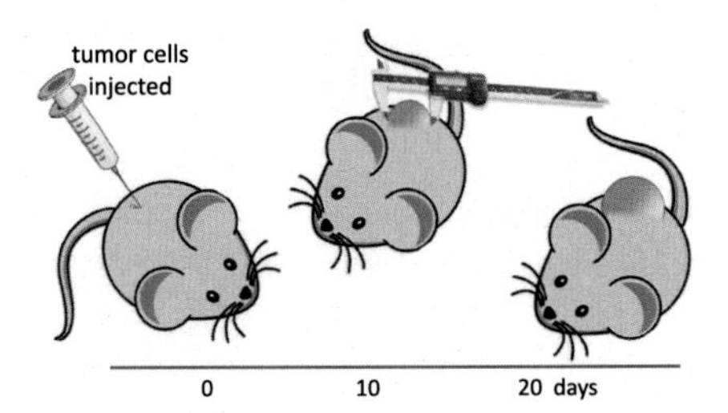

FIGURE 3.5

Tumor cells are injected into mice and allowed to grow. Calipers are used to measure tumor size without harming the mouse.

Table 3.3 Lung tumor data, extracted from Fig. 2 in [4].

Days	Volume (mm^3)
0	1
6	23
11	137
12	189
13	270
14	287
15	383
18	729
19	977
20	1070
21	1312

Exercise 3.18

Plot the data and describe the growth of the tumor over time.

3.5.3 Model formulation

Benzekry et al. [4] considered multiple alternative models for the growth of the tumor. Here we consider four of these proposed models (Table 3.4). How do these models vary? Recall what you know about simple exponential and logistic models from Unit 2. What biological realities are addressed by adding the carrying capacity K in the logistic model? Recall the linear decrease of the relative growth rate with respect to population density, which we saw in lab using the logistic model. Here the logistic model predicts a similar linear decrease in relative tumor growth rate with increasing volume, whereas the Gompertz model is defined by exponential decay in the apparent growth rate in time. An additional model is the power law in which the growth rate is determined by the size of the tumor raised to some power less than one

Table 3.4 Proposed models for the growth of a lung tumor. Assume $V(0) = V_0 = 1$ mm^3.

Model Name	Differential Equation	Solution
Exponential	$\frac{dV}{dt} = aV$	$V(t) = V_0 e^{at}$
Logistic	$\frac{dV}{dt} = aV\left(1 - \frac{V}{K}\right)$	$V(t) = \frac{V_0 K}{(V_0 + (K - V_0)e^{-at})}$
Gompertz	$\frac{dV}{dt} = ae^{-\beta t}V$	$V(t) = V_0 e^{\frac{a}{\beta}(1-e^{-\beta t})}$
Power Law	$\frac{dV}{dt} = aV^{\gamma}$	$V(t) = \left((1-\gamma)at + V_0^{1-\gamma}\right)^{1/(1-\gamma)}$

$(0 < \gamma < 1)$ based on the well-established allometric relationship between metabolic rate (and thus perhaps potential for growth) and animal body size as we saw in Section 1.2.7 (see Fig. 1.7).

Exercise 3.19

Discuss the differences in long-term behavior among the models. (Note that you can determine the long-term behavior by considering what happens to the solution $V(t)$ as t becomes large, that is, taking the limit of each solution as $t \to \infty$.) Be sure to relate the growth trajectories predicted by each model to something you think might limit or enhance tumor growth with size.

3.5.4 Parameter estimation

Your job is to first determine the best-fit parameters using `nls` for each of the models in Table 3.4 and then to compare how well each model represents the existing data using the corrected Akaike information criterion (AICc).

Exercise 3.20

Using the data in Table 3.3, follow the steps below to find the best-fit parameter a for the exponential model in Table 3.4. Recall that to use `nls`, we must have a good initial guess for the parameter. In parts (a) and (b), we use tools learned in the background section to estimate a.

(a) Create a slider for a and plot the model $V(t)$ along with the data. Refer to Listing 3.1 to see an example of using `manipulate`. Report your estimate of a.

(b) Another method to estimate the best-fit parameter is to directly plot the error and visually estimate where the minimum occurs. This also reinforces the concept of nonlinear regression! Refer to Listing 3.3 for an example.
Recall that nonlinear regression finds the parameter values that minimize the sum of the squared residuals given by

$$RSS(p_1, \ldots, p_m) = \sum_{i=1}^{n} (y_i - f(t_i, p_1, \ldots, p_m))^2. \tag{3.9}$$

In this example, we have $n = 10$ data points, the model is given by $V_0 e^{at}$, and a is the unknown parameter. Therefore Eq. (3.9) becomes

$$RSS(a) = \sum_{i=1}^{10} (V_i - V_0 e^{at_i})^2, \tag{3.10}$$

where (t_i, V_i) are the data points. The right side of the equation is now a function of a, and we need to find the value of a that minimizes this function. Write code to evaluate and plot RSS for a range of values for a. Note that the range of RSS is quite large, so it is helpful to plot $\ln(RSS)$ instead of RSS to make it easier to see the minimum. Estimate the value of a that minimizes the logarithm of the sum of the squared residuals. How does this value compare to what you found in part (a)?

(c) Use your estimate from part (a) or (b) as an initial guess in the `nls` command. Refer to Listing 3.2 and Exercise 3.12 for examples using `nls`. What is the best-fit value for a using nonlinear regression? Report your value of a in a table that includes each of the models and associated parameters and then the best fit parameter values based on `nls` as they are determined. (We will be determining the AICc and the ability of each model to forecast, so you may want to plan a master table for all these results.)

(d) Plot the model output and raw data and describe any discrepancies you can visually identify.

Exercise 3.21

Using the data in Table 3.3, follow the steps below to find the best-fit parameters a and K for the logistic model in Table 3.4.

(a) Use the `manipulate` function to create sliders for a and K. Vary these parameters until you find a set of values that produces a reasonable fit to the data.

(b) When we need to find initial guesses for two parameters, sometimes it can be difficult to use sliders to do this; it might be hard to find a combination of the two parameters that fit the data well, or there might be more than one combination that seems to fit the data well. Therefore we can directly plot the sum of squared residual as was done in the previous exercise. However, in that example the sum of the squared residuals was a function of one parameter, and therefore we could plot RSS versus the parameter in a 2-D plot and determine where the minimum occurs. In the logistic model, RSS is a function of two parameters,

$$RSS(a, K) = \sum_{i=1}^{10} \left(V_i - \frac{V_0 K}{V_0 + (K - V_0)e^{-at_i}} \right)^2. \tag{3.11}$$

This means there are two independent variables a and K and one dependent variable RSS. We could consider a range of values for a and K and produce a 3-D plot, but it is easier to visualize using a heat map, as discussed in Section 3.2.4. Use Listing 3.5 as a template to calculate $RSS(a, K)$ for a range of a and K values. Similar to the previous exercise, the range of RSS is quite large, so plotting $\ln(RSS)$ instead of RSS makes it easier to see the minimum. Recall that the package `plotly` has great interactive features that allow you to place your cursor over the plot, and it gives the values of the coordinates (see Fig. 3.6). This will help you find approximately where the global minimum occurs. You can also zoom, which may help you find the minimum. Estimate the values of a and K that minimize RSS. How do these values compare to what you found in part (a)?

(c) Use your estimates from part (a) or (b) as initial guesses in `nls`. What are the best-fit values for a and K using nonlinear regression? Add these to your table.
(d) Plot the model output and raw data and describe any discrepancies you can visually identify.

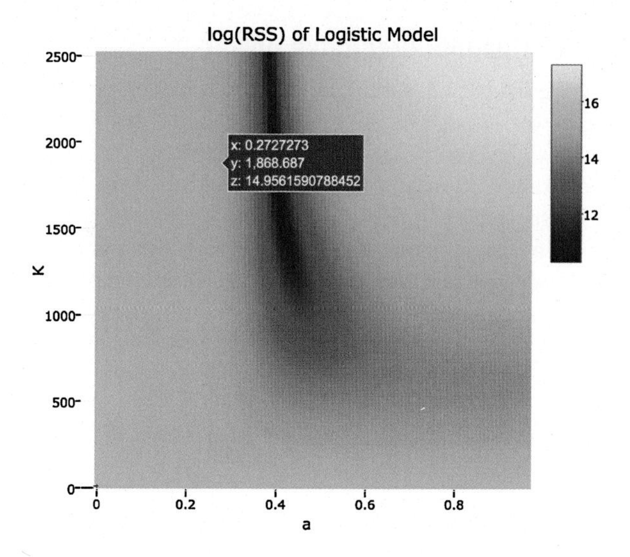

FIGURE 3.6

Plot of $RSS(a, K)$ for the logistic model.

Exercise 3.22

Perform nonlinear regression to find the best-fit parameters for the Gompertz model and power law model. You can either create sliders for your parameters or create a heat map to obtain initial guesses for the parameters.

(a) Report your best-fit values for the parameters in your table so that you can readily compare parameter values among the models. For example, how different are the estimates for a?
(b) On one graph, plot the data and each model and describe any discrepancies you can visually identify.

3.5.5 Model evaluation: descriptive power

There are various ways to measure the descriptive power of each model. Since we are testing alternative models, we will use the AICc to compare them.

Exercise 3.23

Calculate the AICc for each model using the parameters that you reported in Exercise 3.22. (The AICc formula is given in Eq. (3.8). You can also refer to Section 2.4.5 in Case Study 1 where AICc calculations were performed.)

Tabulate the AICc values associated with each model. What conclusions can you make? Do the AICc values support your visual estimation of which model(s) are most suited to this data set?

3.5.6 Model evaluation: predictive power

In the prior section the growth assumptions of each model were tested by assessing the descriptive power of each model against the entire 21-day set of experimental data. These models may also be used to estimate the prior or future course of tumor progression. For example, it may be useful to project backward to know when a tumor started, and it is especially helpful to project likely future tumor growth for clinical assessment or when designing research strategies for antitumor drug therapies. However, it is important to first test the predictive power of each model. There are various ways to measure and compare the predictive power of models. Here we will test the prediction of tumor size at day 21 based on the first five data points by observing the fit to the data. Then we will measure the error using the relative error

$$RE = \left| \frac{V_{10} - V(21)}{V_{10}} \right|,$$

where V_{10} is the tenth data point (i.e., the actual tumor volume at day 21), and $V(21)$ is the model prediction for day 21.

Exercise 3.24

Work through the following steps to test the predictive power of each model. It would be handy to add the relative error for each model to your table.

(a) Using the first five data points in Table 3.3, use nonlinear regression to find the best-fit parameters for each of the four models.
(b) Calculate the volume prediction at day 21 for each model.
(c) Calculate the relative error between the model prediction and actual volume at day 21 for each model.

Exercise 3.25

What conclusions can you make based on your results from Exercises 3.23 and 3.24? In your discussion, include how you reached those conclusions. Are there assumptions leading you to these conclusions? What would you do next with this problem experimentally, in terms of modeling or data analysis? For example, are there model modifications you would suggest?

Exercise 3.26

Further Exploration: Vary the number of data points used to fit the model and then test the predictive power on various data points. If you only had late time points, which models will give you the best estimate of the very early tumor growth rates? You may also want to consider the very realistic situation for which the number of data points is sparse but distributed over the 21-day time-course.

3.6 Case study 3: Predator responses to prey density vary with temperature

Goals: We will explore the Holling functional responses and how these depend on temperature for a pair of invasive freshwater shrimp eating *Daphnia*. Nonlinear least squares will be applied to fit the model to data and derive key parameters, maximal feeding, and attack rates using two models, the Holling disc equation and Rogers random predator equation. Once the parameters are determined at each temperature, we will ask if they fit the expected temperature dependence of biological functions. If we can find a model that fits the temperature dependence of these parameters, the model could be used to predict predation capabilities of each shrimp as a function of temperature.

3.6.1 Background

One of the classic problems in ecology is explaining the relationships between predators and their prey and how these define the numbers of each in an ecosystem. As we have examined previously, population sizes depend on the amount of food available for their own growth and maintenance (e.g., concentration of nutrients in our bacterial broth in the Unit 2 lab), and we have represented this in terms of the carrying capacity of an ecosystem. We also know that population sizes are limited by predators in classic examples, such as foxes eating hares or goats eating grass.

Earlier in this unit, we introduced the Lotka–Volterra model that describes predator–prey interactions given by

$$\begin{aligned} \frac{dx}{dt} &= \alpha x - f(x)y, \\ \frac{dy}{dt} &= \delta f(x)y - \gamma y, \end{aligned} \tag{3.12}$$

where x and y represent the prey and predator populations, respectively, α is the growth rate constant for the prey, δ converts the rate of the consumption of prey to the number of new predators formed, and γ is the intrinsic death rate of the predators. The function $f(x)$ is the *functional response* or the relationship between consumption rate and prey density. The units of $f(x)$ are the number of prey per predator per unit time, and multiplying this by the number of predators y gives the number of prey consumed per unit time.

In Exercise 3.5, we introduced the three fundamental types of functional responses that might describe how a predator responds as the number of prey or prey density increases. A Type I response assumes a simple linear relationship $f(x) = \beta x$, where β is the consumption rate constant. This means that the term $f(x)y$ is written as βxy, which is a simple mass action term. In Exercise 3.5, we also learned about two other typical responses, Holling Type II and Type III responses. Type II responses describe situations in which the rate of prey consumption increases linearly with prey density but ultimately reaches a maximum value with feeding limited not by prey density but perhaps by digestion time. The Type II functional form is given by

$$f(x) = \beta \frac{x}{D+x} = \frac{ax}{1+ahx}. \tag{3.13}$$

As we saw in Exercise 3.5, the first expression $\beta \frac{x}{D+x}$ clearly shows that there is a maximal feeding rate denoted by β. The feeding rate is at half maximum when the number of prey is equal to D. The Type II functional response is often written in the form of the second expression $\frac{ax}{1+ahx}$. Here biologically relevant parameters are included explicitly, where a is the capture efficiency, that is, percent eaten during the search period, and h is the handling time per prey, that is, how long it takes for a predator to complete the job of manipulating and eating the prey. A little algebraic manipulation can help relate the parameters in these two expression; β is equal to $\frac{1}{h}$, suggesting that the rate limiting step at high prey numbers is how fast the predator can process one, and D is $\frac{1}{ah}$, which has units of numbers of prey.

Like the Type II response, Holling's Type III functional response saturates at high prey density, but it takes into account that the predator's ability to catch prey may change as a function of the prey concentration if, for example, there are a limited number of good prey hiding places available or large groups of prey can protect one another. This can be represented as

$$f(x) = \beta \frac{x^b}{D^b + x^b}. \tag{3.14}$$

For $b > 1$, this relationship is a sigmoidal shape characteristic of a Type III response. The parameter b changes the steepness of the curve, and it allows us to generalize how the response depends on prey density.

Experiments can be designed to determine the type of functional response and the corresponding parameter values. Data are typically collected over a fairly short period of time in well-defined environments. The number of prey species is systematically varied over an appropriate range, a fixed number of predators is added, and the number of prey eaten during the experimental time period is recorded. The functional response models can be fit to the data, and the best model can then be selected for the particular experimental conditions.

Exercise 3.27

(a) Draw a picture representing the experiment described and discuss what data are collected. Brainstorm about how the predators might interact with the prey at low, intermediate, and high densities. For a concrete example, think of a fish eating mosquito larvae.

(b) On one graph, sketch a Type I response, Type II response, and Type III response.

(c) Brainstorm about predator–prey examples that might be best described by a Type I response, a Type II response, and finally a Type III response. Talk about which of the underlying components, for example, the capture efficiency or handling time, might change in each circumstance.

(d) Convince yourself that the two expressions for the Holling Type II relationship (the two expressions in Eq. (3.13)) are indeed the same and write β and D in terms of a and h. Explain why $1/h$ represents the maximum feeding rate.

Can predator–prey interactions predict invasive behavior?

There are numerous reasons to study predator–prey systems. Fundamentally, it is interesting to understand the basic ecology quantitatively, not just who eats whom, but what factors significantly control population sizes. Is a predator able to survive with only one prey species in an area? How does the suite of predators alter the potential prey populations? How do external factors such as the landscape itself, temperature, or humidity alter the abilities of the predator to catch and process the prey or the prey to hide or escape? Currently, there are multiple factors disrupting biological communities. Habitat loss changes the capacity of an area to sustain its original group of species (recall our island biogeography case in Unit 2), introduced species disrupt the predator–prey balance evolved in an area over time, and climate change is rapidly altering rainfall, salinity, and temperature patterns, which in turn affect the well-being of the organisms adapted to prior abiotic conditions.

One factor that determines if an introduced species will become invasive is its ability to thrive and dominate in its new environment and the extent to which it is harmful to the population sizes of other species. We saw in the case of the garlic mustard case study in Section 2.5 that its ability to invade is due to its high fecundity and seed survival. Here reproduction is included in the Lotka–Volterra model in Eq. (3.12), but we are now focusing on predators that depend on prey for food. These interactions determine the population of each in their ecosystem. We specifically consider the predator's effects on the prey population starting with the predator's functional response to the prey, that is, feeding rate as a function of prey density.

We will examine the functional responses of two different mysid shrimp predators (shown in Fig. 3.7) as functions of temperature. One of these predators, *Mysis diluviana*, commonly known as the opossum shrimp, is native to the deep, oligotrophic, cold lakes found in Northern Canada, the Great Lakes, and lakes in upstate New York and northern New England. Interestingly, upon introduction to new environments, *M. diluviana* have proven highly invasive, altering the food web by changing the zooplankton composition and thereby the populations of fish that eat them. *Hemimysis anomala*, the bloody red shrimp, is native to the Caspian Sea region and has traveled through Europe starting in the 1950s as introduced fish food. As of 2006, these shrimp appeared in the St. Lawrence Seaway and most of the Great Lakes, likely via cargo ship ballast water.

Both of these freshwater shrimp are vertically migrating predators that eat a wide variety of zooplankton including daphniids and copepods when they are up in the water column at night. They will also eat algae and detritus, if necessary, giving them the flexibility needed to be successful in new environments. There are two critical differences between them. *Mysis diluviana* is larger at 16–25 mm than the 6–13 mm *H. anomala*, which may result in different responses as functions of prey size. In addition, *M. diluviana* have an optimum growth temperature of 8–8.5°C and a maximum

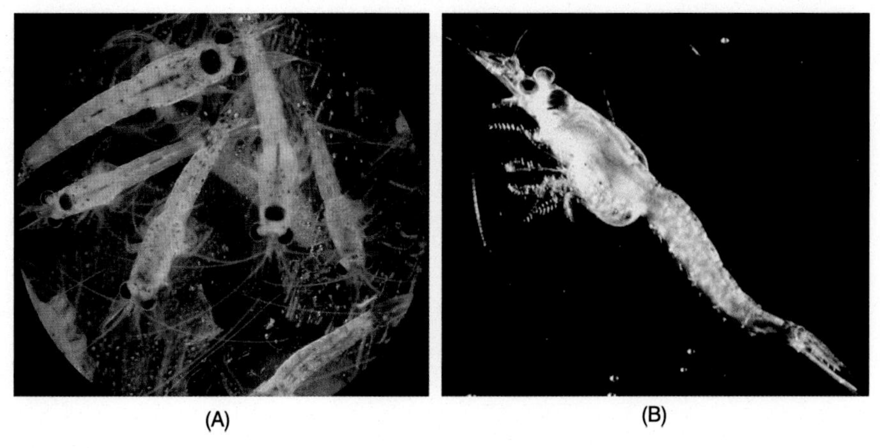

FIGURE 3.7

Hemimysis anomala (A) and *Mysis diluviana* (B). Note the similarities between the two species. *M diluviana*'s size range is significantly larger. Photos were provided by the NOAA Great Lakes Environmental Research Laboratory.

survival temperature of about 25°C [60] in contrast to the optimum growth temperature of 20–22°C and the maximum survival temperature 30–31°C for *H. anomala* [51]. How do their temperature optima and ranges affect their functional response behavior? In the context of global warming, it is likely that decreased ability to capture prey in a new or changing environment may be a key determinant for a species to invade or even survive [26].

3.6.2 Analysis of functional response data: determining the parameters

In these experiments, one shrimp was added to a 170-mL container with a defined number x_0 of *Daphnia pulex* and held in the dark (shrimp are nocturnal feeders) for 12 hours. The remaining *Daphnia* were counted and the number eaten calculated. The data for experiments on *H. anomala* at 5° are plotted in Fig. 3.8. From these data the first step is to determine the form of the functional response (Type I, II, or III) and the interesting parameters, the attack rate and maximum feeding rate.

The first task is to import the data and plot it. Snapshots of the data are shown in Fig. 3.9 and are found in the file FRData.csv [26]. The first column is the species name, the second column is the temperature in degrees Celsius, the third column is the initial number of prey, and the fourth column is the number of prey consumed.

Exercise 3.28

Import the data set FRData.csv. To select the rows of data for only *Hemimysis anomala* at 5°, use the following command:

```
data.hemi.5 = FRdata[FRdata$Mysid.sp=='Hemimysis anomala' & FRdata$temp.C==5, ]
```

Then create a plot similar to Fig. 3.8.

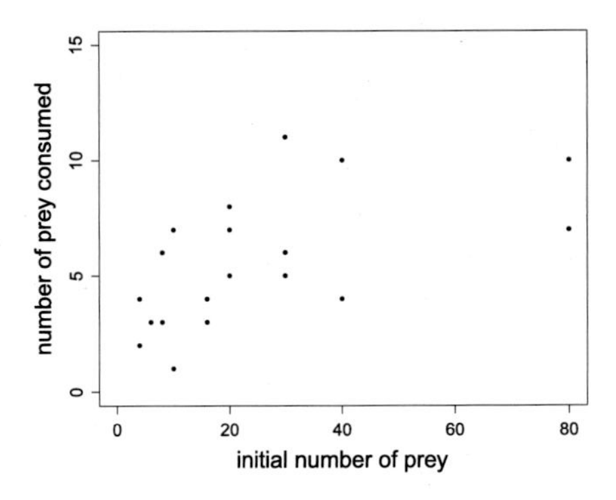

FIGURE 3.8

Functional response of *Hemimysis anomala* to *Daphnia*, its prey at 5° ([26]).

	A	B	C	D
1	Mysid.sp	temp.C	Initial.prey.density	prey.consumed
2	*Hemimysis anomala*	5	4	4
3	*Hemimysis anomala*	5	4	2
4	*Hemimysis anomala*	5	4	4
5	*Hemimysis anomala*	5	6	3
⋮				
150	*Hemimysis anomala*	28	40	4
151	*Hemimysis anomala*	28	40	5
152	*Mysis diluviana*	5	4	4
153	*Mysis diluviana*	5	4	4
⋮				

FIGURE 3.9

Format of raw data.

Since the duration of the experiment is short, we can neglect births and deaths. Therefore the rate of change of the number of prey in Eq. (3.12) can be simplified to

$$\frac{dx}{dt} = -f(x)y. \tag{3.15}$$

Statistical methods can be used to distinguish between Type II and Type III functional responses (logistic regression is typically used to determine functional response type; see [29]). Although we will not perform this step here, these methods can be used to show that the functional response data for the shrimp eating *Daphnia* follow a Type II response. Substituting the Type II response into Eq. (3.15), the rate of change of the prey becomes

$$\frac{dx}{dt} = -\frac{ax}{1+ahx}y, \tag{3.16}$$

where t is in units of hours. If the duration of the experiment is short, then births and deaths do not occur, and the number of predators y remains constant throughout the experiment. Moreover, if there is no depletion of available prey during the experiment (i.e., the prey are replaced after being consumed), then x remains constant throughout the experiment, which we will denote by x_0. Therefore dx/dt is assumed to be constant. In this case, since Eq. (3.16) represents the change in the number of prey per hour, then the number of prey eaten, denoted by x_e, in a period of τ hours is

$$x_e = \frac{ax_0}{1+ahx_0}y\tau, \tag{3.17}$$

which is called the Holling disc equation. The parameter x_0 is equal to the initial number of prey, and it is assumed to be constant if prey are replaced. In our case, at each temperature the experiment lasted 12 hours, and one predator was used in each trial. Therefore the number of prey eaten during the experiment is approximately

$$x_e = \frac{12ax_0}{1+ahx_0} \tag{3.18}$$

for a given number of prey. In these experiments, the eaten *Daphnia* were not replaced, and therefore the rate of change in Eq. (3.16) is not constant since x is decreasing, that is, the prey are depleted. If the experiment is over a very short time period, the Holling disc equation might be a reasonable approximation. First, we will use the Holling disc equation to estimate the attack rate and handling time, and later, we will consider how to incorporate prey depletion and compare the results.

Exercise 3.29

(a) Looking at the data in Fig. 3.8, estimate the attack rate a and the handling time h. It might be helpful to refer to Eq. (3.13) and consider how a and h are related to the maximum number of prey consumed per unit time by one predator and the number of prey at which this rate is at half maximum.

(b) In Exercise 3.28, we imported the data for *H. anomala* at 5° and assigned these to `data.hemi.5`. Fit Eq. (3.18) to these data by using estimates from part (a) in `nls` (refer to Listing 3.2 for a similar example). Save these best-fit parameters a and h as `a.hemi5` and `h.hemi5`, respectively. You will need these values later!

(c) Plot the data, along with the best-fit Holling disc equation.

(d) Fit the Holling disc equation to the data for *Hemimysis anomala* at 10°, 15°, 20°, 24°, and 28°. Plot the data and model at each temperature and report your values of a and h at each temperature.

(e) In the experimental study, the prey were not replaced after being depleted. However, the Holling disc equation does not take into account prey depletion and therefore assumes the rate at which the prey are being consumed is constant. Would the attack rate a and handling time h in the

Holling disc equation underestimate, overestimate, or be the same as the actual values of a and h? Explain your answer.

Consider again the differential equation (3.16), which describes the rate of change of prey. For conditions without prey replacement, the right-hand side is not constant because x is decreasing in time. Using separation of variables, a method learned in a differential equations course or calculus course, we can solve for $x(t)$ and evaluate this at τ, the duration of the experiment, to obtain

$$x_e = x_0\left(1 - e^{ahx_e - \tau y}\right), \tag{3.19}$$

where x_0 is the initial number of prey, and x_e is the total number of prey consumed, which is equal to $x_0 - x(\tau)$. Eq. (3.19) is called the Rogers random predator equation, which takes into account prey depletion. We could fit this model to the data and estimate a and h, but note that x_e appears on both sides of the equation! This is an implicit equation, and it is not possible to isolate x_e on one side. Therefore `nls` cannot be used here. In such a case, we can use alternative methods (one such method will be discussed in Unit 4). For now, in the next exercise, we will utilize `FRAIR`, an R package for fitting and comparing functional responses [54].

Exercise 3.30

(a) Install and load the `frair` package. Then use the following command to fit the Rogers random predator equation to the data for *Hemimysis anomala* at 5°:

```
fit = frair_fit(formula=prey.consumed~Initial.prey.density, data = data.hemi5,
      response = "rogersII", start = list(a = 0.1, h = 0.1), fixed = list(T = 12))
```

The first argument of `frair_fit` is the formula (similar input to `lm`, where the dependent variable comes first, and the independent variable comes second), followed by the name of the data frame where the data are stored, the type of response, the initial guesses for a and h, and any fixed parameters (here the duration of the experiment, denoted by `T` in `frair`, is fixed). You can plot the raw data and overlay the model fit:

```
lines(fit,col="blue")
```

and save the estimated coefficients using similar commands when saving results from `nls`:

```
a.rog.5 = fit$coefficients[["a"]]
h.rog.5 = fit$coefficients[["h"]]
```

(b) Repeat part (a) to find the best-fit parameters of the Rogers equation for the other temperatures.

(c) In Exercise 3.29(d), you predicted how the best-fit values of a and h in the disc equation would compare to values found using a model that assumes prey depletion. Now that you have parameters for the Rogers equation, compare the parameters for the two models. Was your prediction correct?

Exercise 3.31

Repeat Exercises 3.29 and 3.30 for *M. diluviana*.

The `FRAIR` package also has a function that can perform a statistical test to distinguish between Type II and Type III responses (try this using `frair_test(formula = prey.consumed Initial.prey.density, data = data.hemi5)`). In addition, there is a function that can fit the Holling disc equation and other types of functional responses (in `frair_fit`, use `response = "hollingsII"`). However, note that these results may differ slightly from your results using `nls`, which uses a least squares approach. The `frair` package instead uses maximum likelihood estimation (MLE), a popular and robust approach to fitting nonlinear models.

3.6.3 Exploring functional responses as a function of temperature

In the face of global warming and/or introductions of species to areas with new temperature ranges, many scientists are considering the temperature dependence of functional responses as a useful model for predicting species' success. Functional responses are fairly easy to measure, and as we have seen in this case study, these responses can be used to estimate the attack rate, an activity indicator, and maximum feeding rate, an indicator of ecosystem impact.

We would expect both the attack rate and maximum feeding rate to vary with temperature. The mysid shrimp are ectotherms, which means that their body temperatures conform to that of the environment. Digestion rates, muscle function, and neurotransmission are all temperature sensitive to varying extents.

How do you think the attack rate and maximum feeding rate will vary with temperature? Think about what biological activities are represented by these parameters and sketch these relationships. In the next exercise, you will plot your estimated parameters as functions of temperature to compare how these results compare with your predictions.

Exercise 3.32

(a) Using your results from Exercise 3.30, plot the parameter value a versus temperature for *H. anomala*. Discuss your results.
(b) Repeat part (a) but instead plot $1/h$ versus temperature.

In general, the rates of biological processes increase with temperature to a maximum or optimal value (T_{opt}) and then begin to drop off fairly precipitously with warmer temperatures until a critical temperature at which function ceases. As noted, the temperature ranges of *M. diluviana* and *H. anomala* overlap at low temperatures, but only *H. anomala* survives above 20°C.

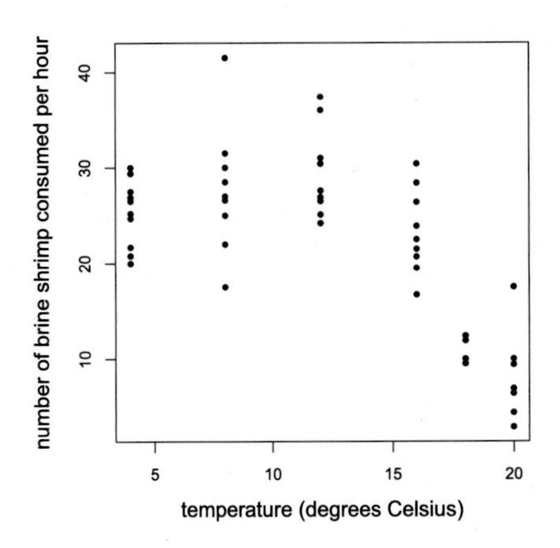

FIGURE 3.10

Feeding rates of *M. diluviana* (formerly called *M. relicta*). Brine shrimp nauplii were consumed over a four-hour period. Each point represents results for one animal. Data extracted from Fig. 2 in [60].

Can we model the temperature dependence of the attack rate a and the maximum feeding rate $1/h$ for these two shrimp? These parameters typically show a hump-shaped relationship with temperature. One way to represent this type of relationship is with a Gaussian function

$$p(T) = p_{\max} e^{-(T-T_{\text{opt}})^2/(2s^2)}, \tag{3.20}$$

where T_{opt} is the temperature at which the function $p(T)$ reaches its highest value of $p_{\max}$, and s is a measure of the breadth of the function. The variable p here could represent attack rate a or maximum feeding rate $1/h$.

Exercise 3.33

(a) In Exercise 3.32, you plotted your estimates for the attack rate a as a function of temperature for *H. anomala*. Fit the Gaussian function in Eq. (3.20) to these data (letting p represent the attack rate). Plot the data and model fit on the same graph and report the estimate of T_{opt}.

(b) Repeat part (a) but now fit the Gaussian function to your previous estimates of maximum feeding rate $1/h$ as a function of temperature.

(c) How could your results from parts (a) and (b) be used to understand the ecological impacts of this invasive species in a changing climate and inform management strategies?

(d) A Gaussian function is symmetric about the mean, but many physiological processes vary asymmetrically with temperature. A common model used to represent this relationship is a piecewise function [13]:

$$p(T) = \begin{cases} p_{\max} e^{-\frac{(T-T_{\text{opt}})^2}{2\sigma^2}}, & T \leq T_{\text{opt}}, \\ p_{\max}\left(1 - \frac{T - T_{\text{opt}}}{T - T_{\text{crit}}}\right)^2, & T \geq T_{\text{opt}}. \end{cases} \quad (3.21)$$

Describe the assumptions of this model and how they differ from a Gaussian function over the entire interval. Then compare and contrast this model with the alternative temperature-dependent function presented in the photosynthesis example in Exercise 3.9.

The results of this exercise can help us understand the effect of temperature on various parameters, but we should be cautious when making conclusions based on fitting a model with three parameters to only six data points. Small data sets can lead to overfitting, which means that there are more parameters than can be justified by the data. This can lead to a model that fits too closely to a particular set of data and therefore cannot reliably fit additional data. Another danger in using small data sets is that any outliers will highly influence the model fit. Although sometimes it is impossible to avoid using a small data set and insights can still be gained, any conclusions from the model analysis should be reported with caution.

The functional response data for *M. diluviana* was collected at only three temperatures. The `nls` function will not even work in this case since the number of data points is equal to the number of unknown parameters! However, there are similar data from a previous study that can be used to estimate T_{opt} for the feeding rate.

Exercise 3.34

(a) Using your results for *M. diluviana* in Exercise 3.31, plot the feeding rate $1/h$ as a function of temperature. Does there seem to be an optimal temperature? What is your guess for this value?

(b) Import the data from Fig. 3.10, which is stored in MysisData.csv. Fit the Gaussian model in Eq. (3.20) to the data. What is the optimal temperature for the feeding rate?

(c) How does the feeding rate optimal temperature for *M. diluviana* compare to that of *H. anomala*? What further data would you collect to be more confident in your conclusions? How would you use these results to help assess the risk due to these invasive species?

3.7 Wet lab: enzyme kinetics of catechol oxidase

Goals: We will derive the Michaelis–Menten equation from the principles of mass action kinetics applied to an enzyme reaction system. We will extract an enzyme and measure reaction rates to yield a data set suitable for using nonlinear regression to determine the Michaelis–Menten kinetic parameters K_M and V_{max}. Linear regression will also be used on common transformations of the data to estimate K_M and V_{max}, and these results will be compared to those from using nonlinear regression. Finally, we will use these experimental and computational tools to discriminate between a competitive and noncompetitive inhibitor.

3.7.1 Overview of activities

We will explore the properties of chemical reactions that are catalyzed by special proteins known as enzymes. Then we will use a set of assumptions to derive the Michaelis–Menten model of enzyme kinetics and examine how inhibitors might affect the model mathematically and experimentally. We will determine the kinetic parameters of catechol oxidase, the enzyme responsible for browning in fruits and vegetables upon exposure to oxygen. To do this, we will prepare a crude extract containing catechol oxidase from apples and add it to multiple concentrations of the substrate catechol. Appearance of the brown product will be detected by the increase in solution absorbance at multiple time points to determine initial reaction rates V_0 at each catechol concentration. We will find the parameters K_M and V_{max} for catechol oxidase using both linear and nonlinear regression methods to fit the Michaelis–Menten model. Next, we will manipulate the experimental system by adding enzyme inhibitors and determine their effects on the reaction rates and model parameters as a means of understanding their modes of action.

3.7.2 Introduction to enzyme catalyzed reaction kinetics

In Section 3.2.1, we learned how to write the rate of change of product (or reactant) concentration over time using the law of mass action. For example, consider a typical chemical reaction expressed as

$$A + B \underset{k_{-1}}{\overset{k_1}{\rightleftharpoons}} C,$$

where A and B are two chemicals or reactants that can react together to form a third chemical or product C. The parameters k_1 and k_{-1} are rate constants similar to those we have seen for other processes such as movement between two compartments. The net rate of product formation is the forward reaction minus the reverse reaction:

$$\frac{d[C]}{dt} = k_1[A][B] - k_{-1}[C]. \tag{3.22}$$

Suppose we start with some initial concentration of reactants and no product, C. Initially, the concentration of C would increase over time (because the rate of

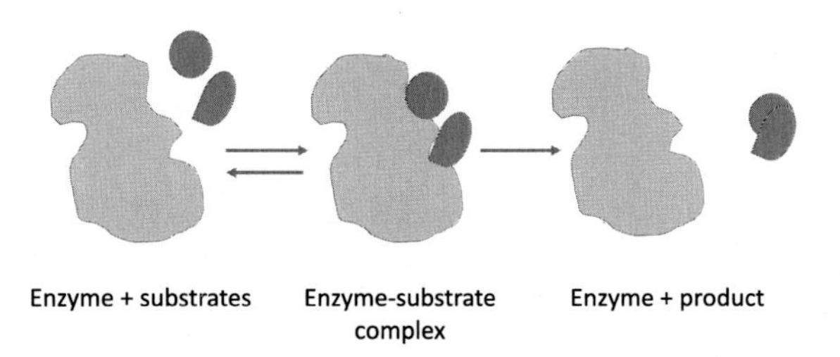

FIGURE 3.11

Diagram of enzyme-substrate binding and catalysis.

the forward reaction, $k_1[A][B]$, would be large compared to the rate of the reverse reaction, $k_{-1}[C]$) until eventually the system reaches equilibrium, that is, the forward and reverse reaction rates become equal.

To tease out the rate constants for a reaction, most laboratory experiments are done far from equilibrium, usually with no product in the reaction mixture, and observed over a short period of time well before equilibrium is reached. Catalyzed reactions occur via a sequence of steps. First, the reactants bind to a catalyst as described by Eq. (3.22), and this rate depends on the concentrations of reactants. Once bound, the reactants undergo chemical change (another rate), and the resulting product is released leaving the catalyst unchanged.

Enzymes are protein catalysts found at low concentrations in vivo. They are critical for most biological processes. Enzymes increase reaction rates and provide specificity by selective substrate binding at specific sites. (Note that when talking about enzymes, the reactants are called substrates. For further reference, see [27,62].) The term *enzyme kinetics* describes how an enzyme catalyzed reaction behaves as a function of substrate concentrations in much the same way as predation rate depends on prey concentrations (see the predator–prey case study). (In this case the units for concentration are in moles of chemical per liter of solution; a mole is a huge number of molecules (6.022×10^{23} molecules per mole), making this an easy way to represent and compare different chemical concentrations.) The reaction is a sequence of steps. An enzyme E binds to a substrate S, forming an enzyme-substrate complex ES, and facilitates a reaction often by changing the conformation of the substrate or its local chemical environment to make it more amenable to the reaction (Fig. 3.11). The reaction occurs, and the enzyme releases the product P. The enzyme can then bind to another substrate. This process is represented by

$$E + S \underset{k_{-1}}{\overset{k_1}{\rightleftharpoons}} ES \underset{k_{-2}}{\overset{k_2}{\rightleftharpoons}} E + P.$$

The concentrations of enzyme, substrate, complex, and product will change over time as the reaction proceeds. When the solutions of the enzyme and substrate are well mixed, the concentration of the complex quickly reaches a quasi-steady state, that is, the complex concentration remains almost constant ($\frac{d[ES]}{dt} \approx 0$). We also observe that the product concentration increases almost linearly at early time points,

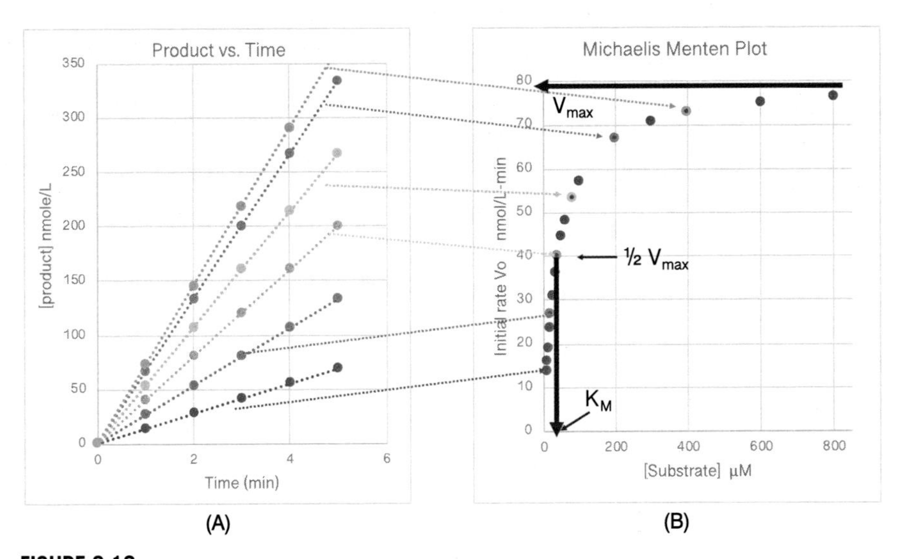

FIGURE 3.12

(A) Time course of product concentration increase for different values of initial substrate concentration. (B) The initial rates of product/time plotted for each concentration of substrate generates the Michaelis–Menten plot.

and we denote this initial reaction velocity by V_0. The product formation rate slows as the substrate is depleted, and finally the substrates and products reach their respective equilibrium concentrations. If we decrease the initial amount of substrate, then the initial reaction velocity will decrease (Fig. 3.12A). Plotting the values of V_0 versus initial substrate concentration $[S]$ results in a curve like that shown in Fig. 3.12B, which is called a Michaelis–Menten plot.

By inspecting the two-step reaction, we can infer that the rates of enzyme-catalyzed reactions are nonlinear with substrate concentration, initially increasing and then saturating, as substrate concentration reaches a value where virtually all of the enzyme is in the ES complex. Since the enzyme is a protein, the reaction rate may be altered by anything that alters protein structure, thus altering substrate binding kinetics (the first part of the reaction) or product formation rates (the second part of the reaction). We typically describe enzyme kinetics using two values, $V_{\max}$ and K_M. $V_{\max}$ describes the upper limit of reaction velocity under a given set of conditions (temperature, pH, amount of enzyme), and K_M is a measure of affinity roughly related to the dissociation constant between S and E.

With the enzyme concentration fixed, the reaction rate at low concentrations of substrate is limited by the probability of productive collisions between the substrate and enzyme and is thus dependent on substrate concentration. At high concentrations of substrate, the reaction is limited by the enzyme turnover rate, that is, the time required for the substrate to be converted to product and its release. Once the product is released, the enzyme can bind to another substrate. $V_{\max}$ is the number n copies of enzyme in the system multiplied by the enzyme turnover rate k_2. K_M is formally defined as the concentration of substrate needed for the initial rate of reaction to be $V_{max}/2$. K_M can be thought of as an inverse measure of the affinity of the substrate for the enzyme. If the substrate-enzyme interactions are fast enough to approach equilibrium, then K_M is close to the dissociation constant between substrate and enzyme expressed as K_S. A low K_M indicates that the binding between the enzyme and substrate is strong; therefore little substrate is needed to reach $V_{max}/2$. If K_M is high, then the affinity is low, and a high substrate concentration is needed to reach $V_{max}/2$. K_M may be expressed in terms of the rate constants relating the interaction between the substrate and enzyme, that is,

$$K_M = \frac{k_{-1} + k_2}{k_1}. \tag{3.23}$$

The first mathematical breakthrough in the study of enzyme kinetics occurred in 1913 when Leonor Michaelis and Maud Menten used differential equations describing each aspect of the reaction with some reasonable assumptions to formulate in a single equation the relationship between substrate concentrations and initial reaction velocities (a useful overall reference is [62]). This equation revolutionized the study of enzymes and enzyme kinetics laying the foundation for various algebraic transformations that are still useful today even though we can access much more detail experimentally [76]. The Michaelis–Menten equation describes the kinetic data seen in Fig. 3.12 and is given by

$$V_0 = V_{max} \frac{[S]}{K_M + [S]}, \tag{3.24}$$

where V_0 is the initial reaction velocity at a given substrate concentration $[S]$ (Fig. 3.12A). Note that the rate constants k_1, k_{-1}, and k_2 are folded into a single parameter K_M, given in Eq. (3.23), to simplify the equation. Before we step back and use the basic rate equations and key assumptions used to derive this famous equation, recall that the Michaelis–Menten equation is of the same form as the functions you explored in Exercises 3.5(b) and 3.7(a).

Exercise 3.35

It may be helpful to relate the Holling Type II and Michaelis–Menten relationships explicitly.

(a) If we were to draw an analogy between enzyme-substrate and predator–prey interactions, which role does the enzyme play? Which role does the substrate play? Explain.

(b) Compare the parameters K_M and V_{max} with β and D as defined in Exercises 3.5(b) and 3.7(a). Include the physical interpretation and units of each parameter.

3.7.3 Deriving the model

Two critical assumptions are needed to derive the velocity equation. Michaelis and Menten used the rapid equilibrium assumption stating that the first step in the reaction is fast compared to the second one, so that ES rapidly reaches a steady-state concentration, and therefore it can be approximated as a constant. We will use this assumption to obtain K_M as described earlier (Eq. (3.23)). The second assumption is that the product concentration remains so low that it never binds to the enzyme and the back reaction does not occur. This latter assumption can be met by careful experimental design to ensure rates are measured before significant product is formed. We are also folding together the conversion rate of the substrate to product and the product release rate from the enzyme. This allows us to write the reaction in a simpler form given by

$$E + S \underset{k_{-1}}{\overset{k_1}{\rightleftharpoons}} ES \xrightarrow{k_2} E + P. \tag{3.25}$$

Exercise 3.36

Starting from the reaction in Eq. (3.25) and using the assumptions stated in the previous paragraph, we will derive the relationship between the initial reaction velocity and substrate concentration, $V_0 = V_{\max}[S]/(K_M + [S])$. Note that early values of $d[P]/dt$ correspond to the initial reaction velocity V_0.

(a) Using the law of mass action, the differential equation $\frac{d[P]}{dt} = k_2[ES]$ represents the rate of product formation. Write a differential equation for the rate of change of complex.

(b) Recall that $[E]$ and $[ES]$ change with time very early in the reaction. Explain why

$$\frac{d[E]}{dt} + \frac{d[ES]}{dt} = 0. \tag{3.26}$$

(c) Initially ($t = 0$) only substrate and enzyme are present at concentrations $[S]_0$ and $[E]_0$, respectively. Explain why Eq. (3.26) implies $[E] + [ES] = [E]_0$, where $[E]_0$ is a constant equal to the total amount of enzyme in the system. Note we can rearrange this as

$$[E] = [E]_0 - [ES]. \tag{3.27}$$

(d) Now we can invoke the steady-state assumption. Fortunately, enzymes are effective at very low concentrations, 100-fold or less than that of the substrate. Therefore we can assume that $[E]_0$ is small. If $[S] >> [E]$, then very shortly after mixing E and S, a steady state will be established in which the concentration of ES remains essentially constant with time. This is the steady-state assumption $d[ES]/dt \approx 0$, which allows us to find an expression for $[ES]$. In part (a), you found an equation for $d[ES]/dt$. Using this equation, substitute your expression for $[E]$ given in Eq. (3.27), set $d[ES]/dt$ to zero, and solve for $[ES]$ to obtain

$$[ES] = \frac{[E]_0[S]}{\frac{k_{-1}+k_2}{k_1} + [S]}. \tag{3.28}$$

(e) Substitute the expression for $[ES]$ in Eq. (3.28) into the equation $d[P]/dt = k_2[ES]$ (where we now denote $d[P]/dt$ by V_0 since we have assumed that we are at the start of the reaction before much product has formed). Let $V_{\max} = k_2[E]_0$ and $K_M = (k_{-1} + k_2)/k_1$ to obtain

$$V_0 = \frac{V_{\max}[S]}{K_M + [S]}. \tag{3.29}$$

This is the velocity immediately after initiating the reaction and before significant concentrations of product are formed or substrate has been deleted.

3.7.4 Estimating K_M and V_{max}

Assuming that we have experimental data on product formation over time for multiple substrate concentrations, how do we use these data to estimate the kinetic parameters K_M and V_{max}? Here we will use sample data to practice estimating kinetic parameters before starting the wet lab.

Suppose tubes containing different concentrations of the substrate ethanol were incubated with 0.17 µg/mL of the yeast alcohol dehydrogenase, a 141 kDa enzyme, in a 3 mL reaction mixture. (In other words, $[E]_0$ was 1.18×10^{-9} M. Why is this important?) The reaction product concentration was dctcrmined at 30 second intervals. Raw data are given in Table 3.5 and plotted in Fig. 3.13A. Estimates of the initial velocities V_0 as µmol product formed per minute for each substrate concentration are also given in Table 3.5 and plotted in Fig. 3.13B. The Michaelis–Menten curve $V_0 = V_{max}[S]/(K_M + [S])$ should describe the data in Fig. 3.13B. To estimate the kinetic parameters, we will use nonlinear regression to find values of K_M and V_{max} that minimize the error between the Michaelis–Menten curve and the initial velocity data.

Table 3.5 Averaged data for the enzyme alcohol dehydrogenase converting ethanol and NAD^+ to acetaldehyde and $NADH$. Data were collected by students in ChBi227 at St. Olaf College.

	Product in µmol/L at:				Estimated velocity
Substrate (mM)	*0 sec*	*30 sec*	*60 sec*	*90 sec*	V_0 *(µmol/L-min)*
2.6	0	0.57	1.45	1.70	1.14
3.3	0	0.79	1.56	2.33	1.58
5.3	0	1.04	2.10	3.16	2.08
11	0	1.59	3.2	4.76	3.18
24	0	2.39	4.78	7.16	4.78
33	0	2.81	5.59	8.41	5.62
67	0	3.31	6.63	9.94	6.62
100	0	4.17	8.34	12.5	8.34
133	0	4.23	8.46	12.6	8.46

Exercise 3.37

Thought questions: Did the experimenter in this example meet the two key assumptions necessary for applying the Michaelis–Menten model? How do you know? Why did the experimenter select these particular concentrations of substrate? Would it have been a problem to use a maximum substrate concentration of say 20 mM?

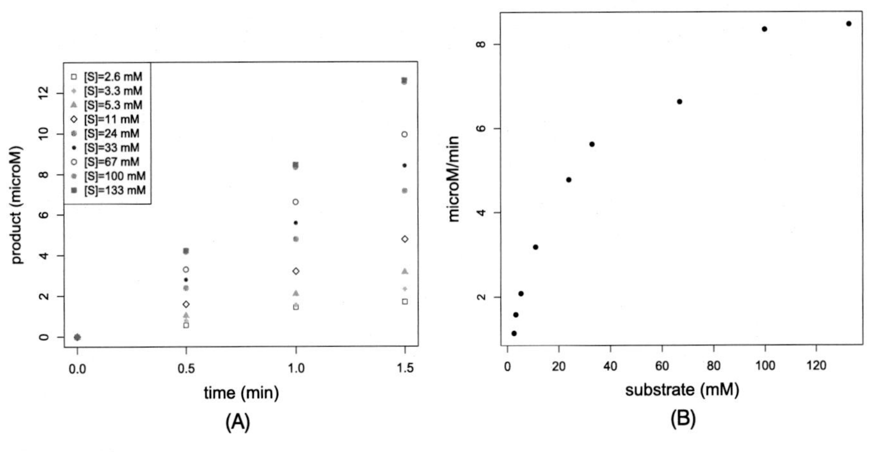

FIGURE 3.13

(A) Raw data given in Table 3.5. (B) Initial reaction velocity versus substrate concentration using values from Table 3.5.

Exercise 3.38

(a) Explain how Figs. 3.13A and 3.13B are related.

(b) The initial velocities were estimated and given in Table 3.5. Suggest how we might go about estimating these initial velocities. Then visually estimate V_{max} and K_M from Fig. 3.13B.

(c) Input the data for $[S]$ and V_0 from Table 3.5 into RStudio. Use `nls` to find estimates of K_M and V_{max}.

In the past, nonlinear regression analysis was not always an option because it was often difficult to access the computational power to perform this analysis. Therefore it was common to transform the dependence between the two variables into a linear relationship. For example, the Michaelis–Menten relationship given in Eq. (3.29) can be rearranged to obtain the Lineweaver–Burk linear transformation

$$\frac{1}{V_0} = \frac{K_M}{V_{max}} \frac{1}{[S]} + \frac{1}{V_{max}}. \tag{3.30}$$

Exercise 3.39

(a) Take the reciprocal of both sides of the Michaelis–Menten equation to derive the Lineweaver–Burk equation.

(b) Using the vectors for V_0 and $[S]$ that you created in Exercise 3.38, create vectors for $1/V_0$ and $1/[S]$. Then, perform linear regression on $1/V_0$ versus $1/[S]$. Reproduce Fig. 3.14A and report the slope and intercept of the best-fit line.

(c) Use the slope and intercept found in part (b) to estimate K_M and V_{max}.

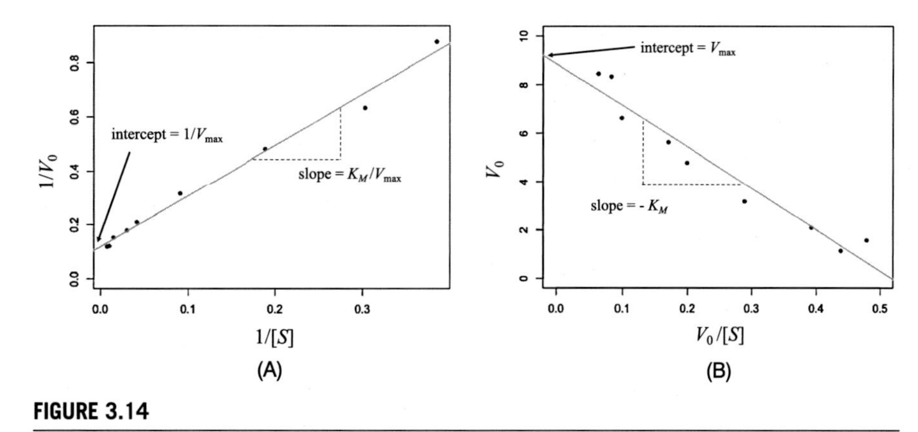

FIGURE 3.14

(A) Lineweaver–Burk plot with best-fit line. (B) Eadie–Hofstee plot with best-fit line.

Another common linear transformation of the Michaelis–Menten equation is the Eadie–Hofstee plot given by

$$V_0 = -K_M \frac{V_0}{[S]} + V_{\max}. \tag{3.31}$$

Exercise 3.40

(a) Rearrange the Michaelis–Menten equation to derive the Eadie–Hofstee equation.

(b) In RStudio, perform linear regression on V_0 versus $V_0/[S]$. Reproduce Fig. 3.14B and report the slope and intercept of the best-fit line.

(c) Use the slope and intercept found in part (b) to find estimates for K_M and $V_{\max}$.

Although both the Lineweaver–Burk and Eadie–Hofstee linear transformations allow us to easily estimate kinetic parameters by drawing a line through the data, these transformations should be used with caution! In the Lineweaver–Burk plot, small errors in V_0 are enlarged when the reciprocal is taken, especially at the lowest rates attained at low substrate concentrations. (Which parameter will be most affected?) In the Eadie–Hofstee plot, both variables V_0 and $V_0/[S]$ are subject to error. Thus the errors in the independent and dependent variables are not independent, as is required for linear regression analyses. Although nonlinear regression analysis is the preferred method to obtain kinetic parameters, linear transformations are commonly seen and can be useful tools for determining types of enzyme inhibition and for diagnosing atypical kinetics.

3.7.5 Our enzyme: catechol oxidase

We will study the kinetics of catechol oxidase, an enzyme responsible for browning in fruits and vegetables [3,37,59]. Catechol oxidase is a protein in the class of

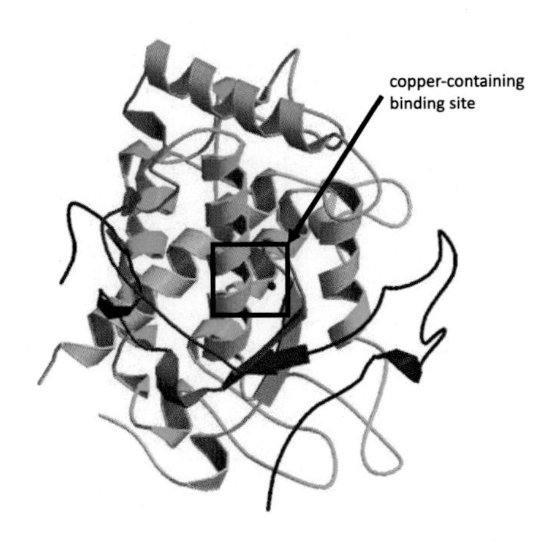

FIGURE 3.15

The protein structure of catechol oxidase: note the copper-containing substrate binding site. Image from the RCSB PDB (rcsb.org) of PDB ID 1BT1 [34].

polyphenol oxidases, a group of copper-containing enzymes that oxidize o-diphenols to form o-quinones, converting hydroxyl (-OH) groups to carbonyl (=O) groups by reducing oxygen with the help of a copper atom [34] (see Fig. 3.15).

Although catechol oxidase is found in animals, fungi, and bacteria, it is most prevalent in plant tissues. Catechol oxidase is responsible for browning in apples and other fruits and vegetables upon exposure to oxygen. The specific physiological benefit of catechol oxidase is currently unknown, although scientists believe that the brown pigments produced may protect the plant tissue from bacterial or fungal growth or possibly insect herbivores. Whereas catechol oxidase and other polyphenol oxidases are crucial in the production of certain products such as tea, prunes, and raisins, enzymatic browning greatly reduces the commercial value of fresh produce. Therefore catechol oxidase and its inhibition are of strong interest to the produce industry. Indeed, catechol oxidase has been removed genetically from some new breeds of apples [5].

Catechol oxidase oxidizes catechol in the presence of oxygen to form benzoquinone (Fig. 3.16). Catechol binds to catechol oxidase's copper-containing binding site to form the enzyme-substrate complex [31], [34]. The product released is o-benzoquinone, a pale yellow compound, which rapidly polymerizes to form a dark brown product. If reaction conditions are right, then the enzymatic steps are rate limiting so that the production rate of dark pigment equals the rate of reaction. Therefore the reaction rate is directly proportional to the increase in absorbance by the dark pigment and can be measured with a spectrophotometer [3].

→ *Note: there are TWO substrates in this reaction, but the Michaelis–Menten model assumes only one. We solve this experimentally by ensuring that O_2 is always present in huge excess so that the substrate of interest, catechol, is reacting with an existing complex of oxygen+catechol oxidase.*

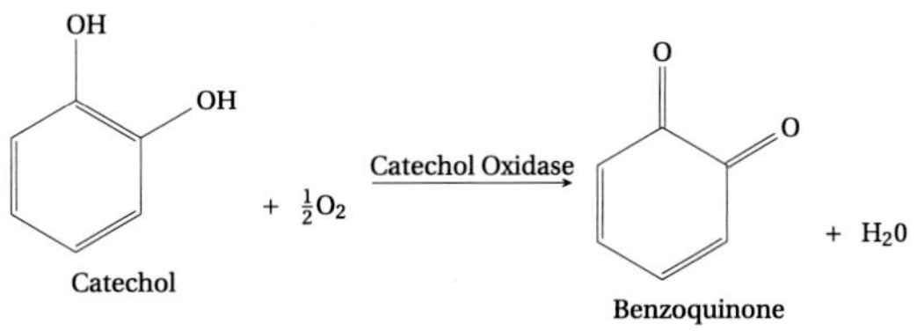

Catechol + $\frac{1}{2}O_2$ —Catechol Oxidase→ Benzoquinone + H_2O

FIGURE 3.16

The catechol oxidase-catalyzed oxidation of catechol to form o-benzoquinone. Benzoquinone rapidly polymerizes to form dark colored pigments.

3.7.6 Experiment: collecting initial rates for the Michaelis–Menten model

Overview of the procedure

We will prepare a crude extract from apples containing the active enzyme. Reaction mixtures with buffer and substrate (catechol) will be prepared. Approximately eight concentrations of catechol including 0 will be used. Reactions will be initiated by adding the enzyme-containing extract, and absorbance will be measured every minute for 10 to 20 minutes. These data sets should allow us to estimate our parameters K_M and V_{max} using R.

Before coming to lab:

1. Read through the entire protocol and draw a flow chart of your procedure in your notebook.
2. Sketch the expected family of raw data plots generated from the various catechol concentrations and show how these will be converted to initial rates and then generate K_M and V_{max} from a Michaelis–Menten plot of these rates. As always, label the axes using appropriate units.
3. Prepare a data table if collecting rates by hand.
4. Note the catechol concentrations and the volume of enzyme/apple extract that you plan to use.

→ *Safety: Wear gloves and eye protection when working with catechol and inhibitor solutions.*

Materials

- Apples
- Catechol solutions (5–8) prepared for final concentrations ranging from 0 to 20 mM. Note: keep these in a dark bottle protected from light.
- 100 mM Na-phosphate buffer pH 7.0, 1 mM EDTA
- Deionized water
- Ice and ice buckets (one/bench)
- Gloves/goggles
- Knife/peeler
- Mortar and pestle or juicer
- Beakers and centrifuge tubes for collecting juice

- Cheese cloth to filter apple juice
- Low or mid-speed centrifuge at 4°C
- Small tubes for distributing aliquots of juice
- Semimicrocuvettes or 96 well absorbance plates (one for each group)
- Pipettes, 20 μL, 200 μL, 1000 μL
- 100 μL octapette if using 96-well plates
- Pipette tips

Enzyme preparation

1. Peel and dice apples before extracting the juice by crushing (e.g., mortar and pestle or hand-held homogenizer) or macerating with the aid of a juicer, to get approximately 10 mL of apple juice.
2. Filter through 2–4 layers of cheesecloth.
3. Transfer to centrifuge tubes.
4. Centrifuge at 1,500–15,000 × g (typically from 3000 to 10,000 rpm) 10 min at 4°C (If spinning at higher speeds, there will be less particulate matter.)
5. Dilute some of the supernatant 1:3 with deionized water (1 part apple enzyme solution to 3 parts water) (Note: this may vary depending on how much catechol oxidase activity is in your apples.)
6. Keep the apple enzyme extract solution on ice in a capped container.

Running the reactions

The particular volumes used and details for combining buffer, catechol, and apple enzyme extract will depend on the instrumentation you are using to measure absorbance.

The following guidelines are a good place to start.

1. Make sure reactants are at room temperature except for the enzyme extract, which should stay on ice.
2. *To confirm workable reaction conditions for your apple extract, carry out a preliminary experiment at a moderate catechol concentration* (e.g., 4 mM) *following the directions below to determine if the amount of enzyme, its dilution or the reaction time need to be altered to attain useful data sets. Every apple variety is different!*
3. Set up reaction mixtures without the enzyme by combining water, catechol in pH 7 buffer in a 2:10 ratio (e.g., 20 μl and 100 μl if using a 96-well plate or 200 μL and 1 mL if using semimicrocuvettes).
4. Mix well and protect the samples from light.
5. When ready to measure the rates, add a small aliquot of the apple enzyme extract (e.g., 20 μL per well or 200 μL per cuvette), mix, and begin recording absorbance data (e.g., at 30 second or 1 min intervals) at 420 nm for about 10 minutes.
6. Repeat until initial reaction rates (absorbance vs. time) are obtained for all substrate concentrations in individual or class data sets.

Analysis in R

Now that the data are collected, the initial reaction velocity can be calculated for each substrate concentration and used to create a Michaelis–Menten plot. Regression techniques can then be applied to estimate K_M and V_{max}.

Exercise 3.41

Calculating the initial reaction velocities

(a) Enter your data in R and plot your data as absorbance vs. time for each concentration of catechol on the same plot. To help you discern the data, use a different color or point type for each concentration of catechol.

(b) Looking at these data, do you see any challenges for calculating V_0, the value you need for the Michaelis–Menten evaluation? Is the relationship between absorbance and time linear? Explain why or why not. Why might the earliest absorbance readings underestimate the reaction rate? Why will the latest time points also underestimate the initial reaction rate?

(c) If necessary, truncate the data at early and/or late time points to obtain the linear portions of the data with steepest slope. There are many ways to do this. How do you propose to choose the points that will most accurately represent the true initial rate? Discuss assumptions and criteria that you are using. Will you use the same time points for every [catechol]? Why not?

(d) Calculate the initial rates V_0 for each catechol concentration as change in absorbance per unit time.

Exercise 3.42

Estimating the parameters K_M and V_{max}

(a) Plot V_0 vs. catechol (substrate) concentration. Does this Michaelis–Menten curve behave as expected? Visually estimate a plausible value for the two key parameters that we are trying to find, V_{max} and K_M. Report these values and their units as initial guesses, which will be used for carrying out the nonlinear fit.

(b) Use nonlinear regression to fit the Michaelis–Menten equation to the data V_0 versus [catechol] using your initial guesses for the values of V_{max} and K_M.

(c) The Lineweaver–Burk approach to estimating the parameters K_M and V_{max} uses a transformation of the Michaelis–Menten equation by plotting 1/[catechol] vs. $1/V_0$ (Eq. (3.30)). Generate this plot and carry out a linear regression of these transformed data. Report the slope and intercept, then calculate K_M and V_{max} from these values.

(d) For the Eadie–Hofstee representation, plot V_0 versus V_0/S (Eq. (3.31)) and perform a linear regression to generate the slope and intercept from which you can calculate K_M and V_{max}.

Exercise 3.43

Data Summary, Questions, and Conclusions

(a) Create a table to report your four different estimates of K_M and V_{max}, that is, by eye and by each of the three methods used to fit the data.

(b) Plot the data V_0 versus [catechol] and overlay the Michaelis–Menten curve using each set of estimates for V_{max} and K_M that you obtained.

(c) Based on your analysis, what are the advantages and disadvantages of each method? Which plot or method do you think gives the most accurate representation of V_{max} and K_M?

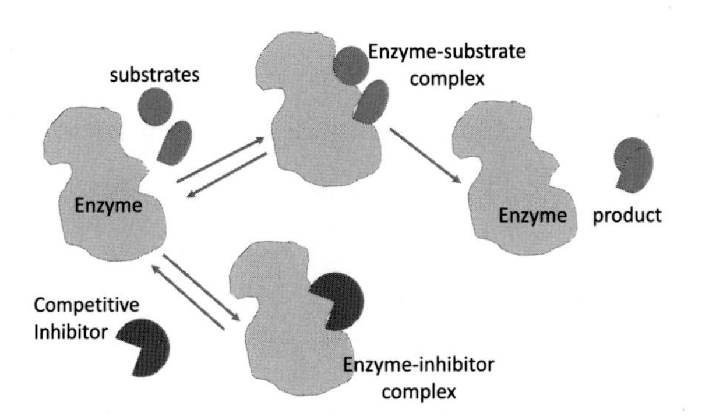

FIGURE 3.17

Diagram of competitive inhibition.

(d) What are the advantages and disadvantages of using spectrophotometry to measure the kinetics of catechol oxidase? Were we able to do experiments that met the assumptions required by the Michaelis–Menten model? What changes would you make in the protocol to improve the accuracy?

(e) Under what circumstances would Michaelis–Menten not be a good model for enzyme activity?

(f) Can you think of other rates in biological systems that could be modeled by the Michaelis–Menten equation?

3.7.7 Effects of inhibitors on enzyme kinetics

It is often advantageous to prevent certain enzymes from working. For scientists, the inhibition of an enzyme gives insight into its structure and mechanism. Many drugs lower enzyme activity, for example, to treat glaucoma, high blood pressure, cardiac insufficiency, cancer, and so on. In the case of our enzyme, catechol oxidase, fruits and vegetables that are damaged during transport in a way that exposes the tissue to oxygen turn brown and quickly become undesirable to customers so food chemistry labs have tested various ways of inhibiting catechol oxidase for decades.

Enzymatic activity can be impeded in many ways. Enzymes are proteins, so all may denature (unfold and lose activity) when exposed to heat. Many enzymes are inactive at extreme pH values for similar reasons. Given that many cooks use acidic lemon juice to prevent apple slices from turning brown, we can surmise that catechol oxidase does not function at low pH. More specific inhibitors are valuable for inhibiting a single type of enzyme without affecting others. This specificity implies that they fit or bind to specific sites on the enzyme.

Specific chemical inhibitors can be classified as **competitive** or **noncompetitive** depending on their mechanism of inhibition. Competitive inhibitors bind to the same binding site as the substrate and thus compete with the substrate (Fig. 3.17). They often have a chemical structure similar to that of the substrate. Noncompetitive inhibitors bind to or disrupt other sites on the enzyme and either effectively remove

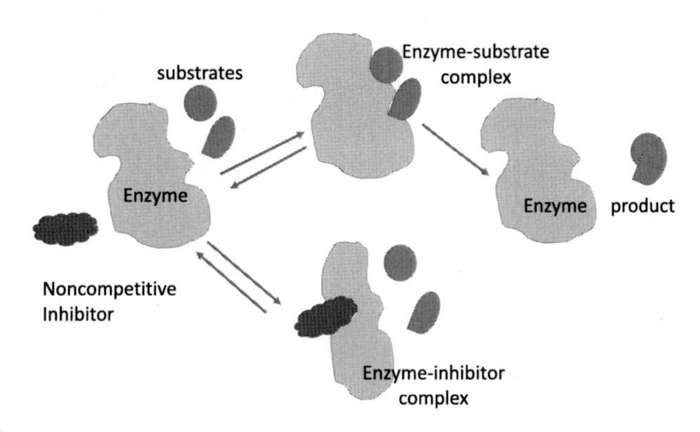

FIGURE 3.18

Diagram of noncompetitive inhibition.

the substrate binding site or alter the enzyme, so that it cannot catalyze the reaction (Fig. 3.18). In doing so a noncompetitive inhibitor essentially lowers the enzyme concentration.

How will these two types of inhibitors affect V_{max} and K_M? When a competitive inhibitor is present, both the inhibitor and substrate will bind to the active site with their respective affinities. However, as the concentration of substrate increases, the substrate will be more likely than the inhibitor to bind at the active site. As a consequence, more substrate than normal is needed to achieve a given reaction velocity in the presence of the competitive inhibitor than in its absence but at very high substrate concentrations, V_0 approaches V_{max}. The K_M appears to increase with little or no change in V_{max}. When a noncompetitive inhibitor is present, the amount of enzyme bound to the inhibitor at equilibrium is effectively removed. Therefore V_{max} decreases, but there is no change in K_M regardless of the amount of substrate present.

In the case of competitive inhibition, the inhibitor and substrate compete for the active site of the enzyme. The original enzyme-mediated reaction now has an additional reaction:

$$E + S \underset{k_{-1}}{\overset{k_1}{\rightleftharpoons}} ES \overset{k_2}{\longrightarrow} E + P$$

$$E + I_c \underset{k_{-3}}{\overset{k_3}{\rightleftharpoons}} EI_c$$

and the initial rate of reaction is

$$V_0 = \frac{V_{max}[S]}{K_M^{app} + [S]}, \tag{3.32}$$

where $K_M^{app} = K_M(1 + [I_c]/K_I)$ is the apparent K_M, and $K_I = k_{-3}/k_3$ is the inhibitor's dissociation constant. K_M is the same K_M measured in an uninhibited experiment. Note that we use the word *apparent* to indicate that what is measured no longer reflects the true affinity between the substrate and enzyme. We see that the term on the bottom indicates that the apparent K_M in the presence of a competitive inhibitor will be greater than the actual value by a factor of $\left(1 + \frac{[I]}{K_I}\right)$.

For noncompetitive inhibition, two reactions are occurring such that $E_0 = E + ES + EI_{nc}$ at any given time:

$$E + S \underset{k_{-1}}{\overset{k_1}{\rightleftharpoons}} ES \xrightarrow{k_2} E + P$$

$$E + I_{nc} \underset{k_{-3}}{\overset{k_3}{\rightleftharpoons}} EI_{nc}$$

and the initial rate of reaction is

$$V_0 = \frac{V_{max}^{app}[S]}{K_M + [S]}. \tag{3.33}$$

Recall that in the uninhibited experiment, $V_{max} = k_2[E]_0$. Here $V_{max} = k_2[E]_0^{app}$ where $[E]_0^{app} = [E]_0/(1+[I]/K_I)$. Since a noncompetitive inhibitor does not bind at the active site, the reaction scheme has an additional potential enzyme complex with both inhibitor and substrate bound. The substrate will never be able to compete with or overwhelm the effect of the inhibitor, and the net effect is to lower the amount of total enzyme which, in turn, will decrease V_{max}.

3.7.8 Experiment: measuring the effects of two catechol oxidase inhibitors, phenylthiourea and benzoic acid

In this lab, we will compare the effects of the catechol oxidase inhibitors phenylthiourea (PTU) and benzoic acid on the kinetic parameters V_{max} and K_M. We will do this at a single concentration of inhibitor to screen them for the type of inhibition, competitive or noncompetitive.

First, compare the structures of phenylthiourea (PTU) and benzoic acid (BA) to that of the substrate catechol (see Fig. 3.19). Can you guess based on the structures how these two inhibitors might work?

Procedure

1. Again, sketch the protocol you are planning to use to determine the mechanisms by which PTU and benzoic acid inhibit the enzyme catechol oxidase.
2. Draw the three model plots, Michaelis–Menten, Lineweaver–Burk, and Eadie–Hofstee, that you would expect for an uninhibited enzyme, a competitive inhibitor and a noncompetitive inhibitor indicating the predicted change in V_{max} and K_M for each. Be able to explain why you drew them this way.

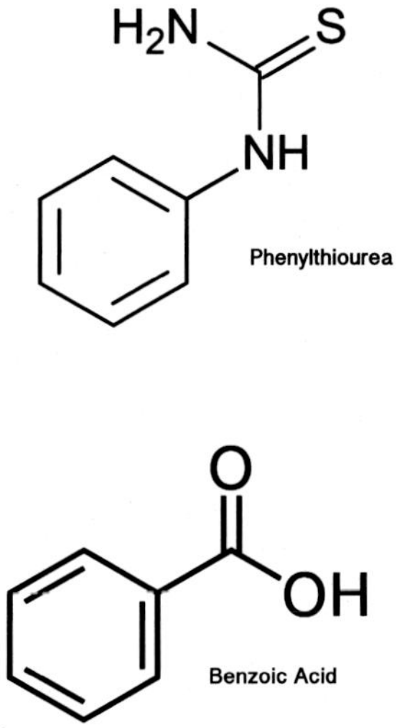

H_2N S NH Phenylthiourea

O OH Benzoic Acid

FIGURE 3.19

Phenylthiourea and benzoic acid are two known inhibitors of catechol oxidase. Compare these structures to that of catechol, also known as 1,2-dihydroxybenzene as shown in Fig. 3.16.

3. Prepare an apple extract/enzyme solution as you did in the first part of this lab using the same tools and materials with the addition of stock solutions of the two inhibitors:
   - 20 mM benzoic acid,
   - 0.4 mM phenylthiourea, and
   - their solvent, 80% ethanol in water.
4. Substitute the volume of water (e.g., 20 μl) for the same volume of one of the inhibitors or solvent. As you set up your reaction mixtures, consider useful controls (e.g., for the effect of the ethanol-based solvent).
5. Prepare the inhibitor and catechol mixtures first. Then, as you did previously, add the enzyme-containing apple extract and rapidly mix just before beginning the absorbance measurements.
6. Collect the data and enter it into R.

Analysis in R

Exercise 3.44

Parameter Estimation

(a) Calculate the initial rates V_0.
(b) Create Michaelis–Menten, Lineweaver–Burk, and Eadie–Hofstee plots for each inhibitor and the control using the R code that you developed in Section 3.7.6.
(c) Calculate V_{max} and K_M for each condition using visual approximation, nonlinear regression, and linear regression on the Lineweaver–Burk and Eadie–Hofstee transformations of the data.

What can we learn from these experiments about how these inhibitors act?

Exercise 3.45

Data Summary and Conclusions

(a) Prepare a table of values of K_M and V_{max} in the absence and presence of inhibitors as estimated by eye and by fitting each of the models.
(b) Are all the estimates equally valid? Why or why not?
(c) For PTU and then benzoic acid, compare the values of K_M and V_{max} to those without the inhibitor.
(d) Given the predicted effects of a noncompetitive vs. a competitive inhibitor on K_M and V_{max}, what are your conclusions about how PTU and benzoic acid work?
(e) What are sources of error for this experiment? Consider the assumptions, the technology, and the theory as possible sources of error.
(f) How would you modify the experimental conditions to find more accurate estimates of the parameter values?

UNIT

Differential equations: numerical solutions, model calibration, and sensitivity analysis

4

Learning outcomes

- Use the *deSolve* package in R to find numerical solutions to systems of differential equations.
- Use the R optimization function `modFit` to find the best-fit parameters in a system of differential equations.
- Perform sensitivity analysis to study how changes in parameters affect the model output.
- Practice using these mathematical tools to develop understanding and new hypotheses to explain complex multivariable biological phenomena, predict future events, and recommend interventions.

4.1 Biological background

Thus far we have modeled a wide variety of self-contained biological events such as the per capita growth rate of a bacterium, the degradation rate of leaves, and the kinetics of a single enzyme reaction. However, we know that most biological events occur in sequences that affect one another. For example, in Unit 2, we studied the interplay between the growth stages of the invasive garlic mustard plant. In the pharmacokinetics case study, we had to consider the drug concentration in multiple compartments and how these affect each other. In this unit, we will explore several biological systems comprised of cascading and often interacting events. Modeling these phenomena typically requires systems of differential equations that can only be solved numerically. Often there are multiple variables and parameters, but only some can be readily measured. Moreover, determining how parameters affect the model can help us understand the system.

The methods introduced in this unit are first applied to the very familiar logistic differential equation, and then we consider a tumor growth model in which the maximum size of the tumor increases as a function of its volume, requiring a system of two differential equations. The background culminates in an example in which we apply these same tools to a system of three equations describing a virus infecting a popula-

Exploring Mathematical Modeling in Biology Through Case Studies and Experimental Activities
https://doi.org/10.1016/B978-0-12-819595-6.00010-4

tion of cells. This model can be used to test strategies for intervening therapeutically or projecting a runaway infection.

On a different scale, we can model an epidemic or behavior of an infectious disease among human or other populations. In the first case study, we introduce the famous SIR model describing how individuals in a population move from being susceptible to a disease, to being infected, and then dead or recovered. We will investigate how the values of model parameters determine whether or not an epidemic occurs. The model allows us to test the role of vaccines, vectors, quarantine, or other measures that might limit the ability of a disease to spread.

In the second case study, we model the growth of prostate cancer cells and the immune response. The immune system responds to a perturbation by a series of steps initiated by recognizing a foreign cell that leads to first initiating and subsequently down-regulating a targeted immune response against the foreign cell. Here we will explore harnessing the immune system as a cancer therapy by developing a model from a vaccine delivered in the dermis to the up- and subsequent down-regulation of an anticancer cell response involving a minimum of six cell types. This example is complicated enough that the model not only helps us design a therapy based on tumor growth data but also reveals the intricacies of the system itself.

In the third case study, we explore quorum sensing, which is the ability of bacteria to "switch" behaviors depending on extracellular conditions, most often the concentration of a signal that all bacteria produce. If the bacterial culture is crowded (i.e., if a quorum is present), then the common signal concentration will increase, and, at a threshold, the bacteria will switch on new characteristics. Typical quorum responses are making a virulence factor, forming a biofilm, or activating a bioluminescent pathway (e.g., *Aliivibrio fischeri*). To do this, the signal molecule, once above threshold, binds to a receptor, which then signals a cascade of events (gene transcription and translation) that permit light formation. Here we combine what we know about bacterial growth curves and introduce a feed-forward mechanism that has a sigmoidal response (Holling Type III-like) to the signal. Quorum sensing models have been especially helpful in developing hypotheses about the design and output of all sorts of complex regulatory systems.

In the lab, we examine a critical homeostatic system, the regulation of our blood glucose concentrations. Blood glucose concentration, like our body temperatures, must be held within very narrow boundaries despite the fact that there are many possible ways to perturb this value. Glucose concentration is itself the signal and, when too high, increases secretion of a hormone, insulin, which initiates a cascade of events to move glucose to cells and store it as glycogen thereby returning the extracellular glucose concentrations to normal. It is very important that since glucose and insulin concentrations have predictable relationships with one another, we only need to measure one of them to estimate the concentration of the other. This is one of the beauties of having a mathematical model: with a set of relatively simple measurements (blood glucose concentrations) and a well-designed experiment, we can predict insulin behavior under several typical biological circumstances without the expense and difficulty of measuring this hormone.

These complex systems with feedback loops typically result in systems of differential equations with nonlinear terms that describe the relationships among the system variables. These systems often do not have analytical solutions, but computational methods can be used to find approximate solutions, which we call numerical solutions. Once the model is formulated and numerical solutions can be found, parameter values need to be defined. If there are values that cannot be found experimentally, it is possible to find estimates given some data set, and we will learn how to calibrate models to find these best-fit parameter values. This method is similar to what we saw in Unit 3, but we no longer have an explicit function that we are fitting to data. The model can then be used to help us understand or manipulate the system (e.g., intervene in an epidemic), but to do this, we should know how robust our model is to changes in the parameters and how each parameter affects the model output. Sensitivity analysis gives us a quantitative method to measure these changes in the output due to changes in parameters.

4.2 Mathematical and R background

To explore more complex biological systems, we will introduce methods to compute numerical solutions to differential equations, find best-fit values for experimentally inaccessible parameters, and understand how each parameter contributes to the overall outcome through sensitivity analysis. To start, we must learn how to solve our models numerically.

4.2.1 Numerical solutions to differential equations

If the goal is understanding biological phenomena, the first step is describing the system with a set of differential equations and then solving the differential equations to see if the model captures the observed or expected dynamics. Systems of differential equations may not have explicit solutions, but we can use computers to solve them numerically. As seen in the previous sections, a simple differential equation such as $y' = ky$ has the analytical solution $y(t) = y_0 e^{kt}$. It is an analytical solution because the variable y can be written explicitly as a function of t. However, since most models describing biological systems tend to be more complicated and involve nonlinear mechanisms, it is often impossible to find an analytical solution. Therefore we must use numerical methods to find approximate solutions to the differential equation.

It is beyond the scope of this text to present the mathematical background and theory of numerical methods. Instead, we will learn how to use the package `deSolve` in R to find numerical solutions to differential equations and understand the output (refer to [66] for a thorough reference on solving differential equations in R). Numerical solutions are *approximations*, and therefore there are always errors in these solutions. Although we can assume that the approximation error is small in the systems studied in this text, we should be aware that numerical methods applied to particular types of equations may result in very large errors. See [23] and [22] for a comprehensive presentation of numerical methods of differential equations.

Example: logistic growth (one variable)

As a first example, we will consider the familiar logistic growth equation

$$\frac{dy}{dt} = ry\left(1 - \frac{y}{K}\right)$$

with $r = 0.5$, $K = 100$, and initial population $y(0) = 10$. To find a numerical solution $y(t)$, the first step is to write a function that we will call `logistic.model` to evaluate the right-hand side of the differential equation, as shown in Listing 4.1.

Listing 4.1 (unit4ex1.R)

```
logistic.model = function(t,y,parms){
  with(as.list(c(y,parms)),{
      dy=r*y*(1-y/K)
      list(dy)
  })
}
```

There are a few new things here that need to be explained! A function is like a miniature program that performs particular tasks and returns the results. Here the function `logistic.model` requires three inputs: time `t`, the variable `y`, and the parameters `parms` (a vector containing parameter values, where in this case, `parms` contains values for `r` and `K`). The function uses these inputs to calculate the derivative at each time and returns these in the form of a list. The R routine `ode` will call this function to solve the differential equation.

The function will be easier to code if we can address the variables and parameters by their names. This is especially helpful when solving large systems of differential equations! The statement `with(as.list(c(y, parameters)), ...)` allows us to use the variable name and parameter names (`r` and `K`).

The basic differential equation solver in R is `ode`, and the format is `out=ode(y0, times, func, parms)`, where `y0` is the initial value of the variable, `times` is a sequence of time points at which the output will be reported, `func` is the function specifying the model (`logistic.model` in this example), and `parms` is the vector of parameter values that will be passed to the function. Therefore we need to specify these inputs before using the `ode` function. Details are shown in Listing 4.2.

The function `ode` is part of the package `deSolve`, which is not included when R is installed. Therefore you must install this by using the command `install.packages("deSolve")`, and this library must be loaded in each new session.

Listing 4.2 (unit4ex2.R)

```
library(deSolve)
y0=c(y=10)    # Initial value of variable
times=seq(0,20,by=0.1)  # Output variable y will be reported at these time points
```

```
parms=c(r=0.5,K=100)    # Defined parameter values

# Solve differential equation
# "out" stores time points and corresponding y values
out=ode(y0,times,logistic.model,parms)

plot(out[,1],out[,2],type="l")  # Plot y versus t
```

The output is stored in the matrix `out` where the first column contains the values of `t` defined by the `time` vector, and the second column contains the values of the variable `y`. Type `head(out)` to view the first six rows of `out`:

```
> head(out)
     time        y
[1,]  0.0 10.00000
[2,]  0.2 10.93669
[3,]  0.4 11.94945
[4,]  0.6 13.04227
[5,]  0.8 14.21890
[6,]  1.0 15.48279
```

Note that the first printed column gives the row number, and it is not part of `out`. We can convert the matrix `out` to a data frame by using the command `out=as.data.frame(out)`. Recall that one advantage of a data frame is that we can call up the columns by names, instead of column number. For example, to plot y versus t, now we can use `plot(out$time,out$y,type="l")`. When we are solving large systems with multiple variables, this will be especially helpful to call up variable names rather than figuring out which column corresponds to which variable!

Note that the solution is a set of points (t, y). Again, this is called a *numerical solution* because it gives discrete values of the approximate solution, which is different from an *analytical solution*, an explicit formula for $y(t)$.

We can combine Listings 4.1 and 4.2 into one function that takes the parameter values and defines and solves the differential equation, shown in Listing 4.3. This format will be useful in later sections.

Numerical Solution to a Differential Equation with One Variable

Below are the commands to compute the numerical solution to the logistic differential equation $y' = ry(1 - y/K)$. The function `logistic` is created to solve the differential equation for a given vector of parameter values, time vector, and initial condition, and the solution is formatted as a data frame. After the function `logistic` is called, the solution is plotted.

Listing 4.3 (unit4ex3.R)

```
# Define function that solve logistic differential equation
logistic = function(parms, times, y0){
  with(as.list(c(parms)),{
```

```
    # Define differential equation
    logistic.model=function(t,y,parms){
        dy = r*y*(1-y/K)
        list(dy)
    }

   # Solve differential equation
   logistic.output = ode(c(y=y0), times, logistic.model, c(r=r,K=K))

   # Convert output to a data frame
   as.data.frame(logistic.output)
   })
}

# Call up logistic function to solve differential equation
# Input to logistic: parms, times, y0
out = logistic(c(r=0.5,K=100), seq(0,40,by=0.1), 10)

# Plot solution y(t)
plot(out$time,out$y,type="l",xlab="t",ylab="y")
```

Now that we have used `deSolve` to solve a very familiar model, the logistic equation, we can apply this method to another more complicated population growth equation in the following exercise.

Exercise 4.1

The spruce budworm is a moth that inhabits northern coniferous forests. The larvae feed on the spruce tree buds and needles. Normally the moths maintain a low density, and their impact on the forest is small. However, approximately every 40 years, their density increases rapidly and results in major defoliation. Ludwig, Jones, and Holling [39] developed a mathematical model to help understand these cycles. A simple model describing the dynamics is given by

$$\frac{dB}{dt} = rB\left(1 - \frac{B}{K}\right) - \frac{B^2}{1+B^2}, \tag{4.1}$$

where $B(t)$ is the budworm density, and t is time, both in arbitrary units.

(a) The second term $\frac{B^2}{1+B^2}$ represents predation due to birds. Note that this is a Type-III functional response (discussed in Exercise 3.5 in Unit 3)! Why might this type of response be used in this model?

(b) If predators were not present, what would happen to the budworm population over time?

(c) Use Listing 4.3 as a template and replace the logistic differential model with Eq. (4.1). Use the parameter values $r = 0.55$ and $K = 6$ and initial condition $B(0) = 0.5$. Solve Eq. (4.1) in R and plot the budworm density over time. You should see the population reach a steady state. How does this steady-state value compare to K? Why is it lower?

(d) To explore the behavior of this model, solve the model for values of K ranging from 1 to 20. (Note: You can create a slider for K using `manipulate`.) Increasing K represents the change in the carrying capacity as the forest grows taller slowly over time. How does the long-term density of budworms change as the forest grows? Do you see a threshold when the population jumps to a high level? How does this relate to the cycles observed in the budworm density?

Now that we are familiar with setting up a single differential equation model and using the `ode` function to find the numerical solution, we will now apply these tools to a system of differential equations. We will do this by examining a new tumor growth model.

Example: tumor growth model (a system of two differential equations)

The models in the previous example and exercises were each a single differential equation with one unknown variable. Most biological systems have interacting parts that change in time, and therefore multiple variables must be included in the model. The method to solve a system of differential equations uses the same commands seen in Listing 4.3. The additional differential equations and initial conditions are easily included, but we must take care that the order of these parts is correct.

In Section 3.5, we tested different models of tumor growth. Here we consider an alternative model that reflects the potential for the tumor carrying capacity to grow as the tumor grows, that is, a dynamic carrying capacity (unlike the models we explored in the case study on lung tumor growth in Section 3.5). The new model is given by two equations,

$$\frac{dV}{dt} = aV\log\left(\frac{K}{V}\right), \tag{4.2}$$

$$\frac{dK}{dt} = bV^{2/3}, \tag{4.3}$$

where $V(t)$ is the total tumor volume, and $K(t)$ is the carrying capacity (maximal volume) at time t. The carrying capacity may depend on the size of the tumor if the blood supply increases with growth. Interesting questions include exploring how altering K as the tumor grows will alter the growth trajectory compared to a fixed value of K. Since it is feasible to interfere with tissue vacularization clinically, quantitative solutions to this model would provide valuable clinical research insights. Before exploring these questions, the system of differential equations must be solved.

Again, we can use Listing 4.3 as a template. Instead of calling the function `logistic`, we will call it `tumorV` and insert Eqs. (4.2) and (4.3). In this example, two initial conditions are required, $V(0) = 1$ mm^3 and $K(0) = 2.6$, and parameter values $a = 0.5$ day^{-1} and $b = 1$ day^{-1}. The time steps will be in days from $t = 0$ to $t = 20$. Try to write the code to solve Eqs. (4.2) and (4.3) and plot $V(t)$ and $K(t)$ on the same graph. Compare your results to Listing 4.4.

Numerical Solution to a System of Differential Equations

Below is the implementation of Eqs. (4.2)–(4.3) with initial conditions $V(0) = 1$ and $K(0) = 2.6$. The function `tumorV.model` returns the derivatives, which must be concatenated in a vector and then packed as a list. The function `tumorV` solves the system of differential equations and returns the solution as a data frame. Note that the order of the derivatives must coincide with the order in which the initial conditions are defined.

Listing 4.4 (unit4ex4.R)

```
tumorV = function(parms, times, V0, K0){
  with(as.list(c(parms)),{
    # Define differential equation
    tumorV.model= function(t,y,parms){
      with(as.list(c(y)),{
        dV = a*V*log(K/V)
        dK = b*V^(2/3)
        list(c(dV,dK))})
    }
    y0=c(V=V0, K=K0)    # Initial conditions passed into function
    out=ode(y0,times,tumorV.model,parms)  # Solve differential equations
    as.data.frame(out)  # Put output into a data frame
  })
}

V0=1; K0=2.6    # Initial conditions
parms=c(a=0.5,b=1)    # Parameter values
out = tumorV(parms, times=seq(0,20,by=1), V0=V0, K0=K0)  # Solve the system

plot(out$time,out$V,type="l",xlab="t",ylab="V")  # Plot V versus t
plot(out$time,out$K,type="l",xlab="t",ylab="K")    # Plot K versus t
```

Run the model and examine the output. Type `head(out)` in the console to make sure that you understand the output! The first column of `out` is t, the second column is V, and the third is K.

4.2.2 Calibration: fitting models to data

In the examples presented thus far in Unit 4, we were given values for the parameters. However, as noted in the Introduction, many parameters are difficult or impossible to obtain experimentally, and one of many reasons to use models is to estimate parameters to test hypotheses (e.g., K_M of an enzyme in the presence or absence of an inhibitor). In Units 2 and 3, we learned that we can find the best-fit parameters by determining the set of parameters that minimize the sum of the squared residuals between the data and model output. In those cases, we had an analytical solution to a differential equation, and therefore we were able to use `nls` to find the parameter values that minimize the sum of the squared residuals. However, the models explored in this Unit have no analytical solutions. Without an explicit solution, we are unable to use nls to implement least squares. The error between the model output and data can still be minimized, but the differential equations will first need to be solved and then compared to the data. A cost function must be defined that calculates the squared residuals, and then an optimization function can be used to minimize the squared residuals. Similar to using nls, initial guesses must be provided, and the optimization can fail if the guesses are not reasonable. Therefore, we will need to consider methods to come up with reasonable guesses for the parameters as input to the optimization function. We will work through these steps using our most familiar model, the logistic differential equation.

Example: calibrating the logistic growth model

Recall that in Unit 3 (Section 3.2.4), we fit the logistic equation to bacterial growth data and used the explicit solution in nls to estimate the parameter values. However, often we do not have an explicit solution. If we have no explicit solution, then we must use the numerical solution to our equation $\frac{dy}{dt} = ry(1 - \frac{y}{K})$ and then compare to the data. We need to solve it using some best-guess parameters and then set up the cost function that calculates the squared residuals for these parameters. Ultimately, the modfit function will use an algorithm to find the parameter values that minimize the squared residuals. How do we do this in R?

First, input the data in R:

Listing 4.5 (unit4ex5.R)

```
t_data=c(0,1,3,4,5,7,8,12,16,18,20,22,27,28)
y_data=c(1.05,1.20,1.56,1.80,1.98,2.28,2.61,3.5,4.7,5.10,5.34,5.49,5.58,5.61)
```

Next, a cost function is defined. This function requires parameter values as input and then solves the differential equation and computes the squared residuals between the model output and data. Since the logistic differential equation was already defined in the function logistic in Listing 4.3, this function can be called up in the cost function. This cost function is defined in Listing 4.6.

Listing 4.6 (unit4ex6.R)

```
# Cost function
logistic.cost = function(parms){
  with (as.list(parms),{
    out = logistic(c(r=r,K=K), t_data, y_data[1]) #  Solve ode for given parameter values
    return((out$y-y_data)^2)  # Return vector of squared residuals
  })
}
```

We will call on the function modfit, which searches the parameter space to find the set of values that minimizes the cost function. Similar to nls, the optimization algorithm in modFit requires initial guesses for the parameter values, and if the guesses are bad, then the optimization routine can fail. There are multiple ways to obtain reasonable guesses for the parameter values. One way is plotting the sum of the squared residuals and trying to find the minimum. For the logistic differential equation, the sum of the squared residuals as a function of r and K could be visualized as a heat map, and then the goal is to estimate where the global minimum occurs. Sometimes it is possible to look at the data itself to make fairly accurate estimates. Here we can make a good initial guess for K by noting that the population density at late times is about 5.6. To obtain an estimate for r, we can use our guess for K and solve for r using data pairs at two intermediate times. Another way to obtain an estimate for r is to use the `manipulate` function, first introduced when finding parameter values using `nls`. An example of using `manipulate` to create sliders with numerically solved differential equations is shown in Listing 4.7.

Listing 4.7 (unit4ex7.R)

```
library(manipulate)
manipulate({
  out = logistic(c(r=r,K=5.6), t_data, y_data[1]) # Solve logistic differential equation
  plot(t_data,y_data) # Plot data
  lines(out$t,out$y,type="l", ylab="y", ylim=c(0.6)) # Overlay model output on data
},
r=slider(0,1,step=0.01))
```

Try running this code and using the slider to adjust the value of r until the curve looks similar to the trend of the data. You should find that if r is about 0.2, the model looks like it fits the data fairly well. The optimization routine `modFit` also allows us to define lower and upper bounds for the unknown parameters, which can increase the efficiency and success of finding optimal parameters. Typically, the parameters are positive values, and therefore it is a good idea to set a lower bound of zero for each.

To carry out the step of actually finding the best fit parameters, we need to install a new library called `FME` (`install.packages("FME")`) and include our initial parameter guesses and their respective lower bounds.

Listing 4.8 (unit4ex8.R)

```
# Optimization resulting in estimates for parameters r and K.
library(FME)
parms = c(r=0.2,K=5.6) # Initial guesses
fit = modFit(f=logistic.cost,p=parms,lower=c(0,0))
summary(fit)
fit$par   # Displays best-fit parameters
```

Note that `fit$par` shows the estimated parameters. If we want to call up specifically r or K, then we can use `fit$par[["r"]]` and `fit$par[["K"]]`. Now that the best-fit parameter estimates have been found, we can compare the model output with the data by plotting both on the same graph.

Listing 4.9 (unit4ex9.R)

```
# Plot data
```

```
plot(t_data,y_data,col='blue', pch=20, xlab="t", ylab='y')

# Solve model using best-fit parameters
mod.pred = logistic(c(r=fit$par[["r"]], K=fit$par[["K"]]), seq(0,30,by=0.1), y_data[1])

# Overlay model output on plot of the data
lines(mod.pred$t,mod.pred$y)
legend("topleft",c("data","model"),lty=c(NA,1),col=c("blue","black"),pch=c(20,NA))
```

Phew! Now, does the output fit your data fairly well? How do the best-fit parameters compare with your initial guesses? Are you ready to apply these methods to the tumor growth model with a dynamic carrying capacity?

Exercise 4.2

Finding Parameters for a Tumor Growth Model with a Dynamic Maximum Size

Recall that the tumor growth model we explored in the previous section was describing tumor size $V(t)$ and maximum volume $K(t)$, and these equations included the parameters a, the tumor growth rate constant, and b, the rate constant for K. Calibrate the model following the steps below.

To begin, we need to input the data. Note that we only have measurements of the tumor volume at particular times.

```
# Tumor volume data
t_tumor = c(0,6,11,12,13,14,15,18,19,20,21)
V_tumor = c(1,23,137,189,270,287,383,729,977,1070,1312)
```

Recall that we already defined the differential equation in Listing 4.4, which we will use here.

(a) Following Listing 4.6, set up the cost function by replacing the logistic function with `tumorV`. Set initial conditions for both variables, $V(0) = 1$ and $K(0) = 2.6$. Also, make sure that you are solving the differential equation at times corresponding to the given data and calculate the residuals between the model output at these times and the tumor volume data.

(b) Use `manipulate` to find reasonable guesses for the parameters a and b using Listing 4.7 as a guide to create sliders for these parameters. Once you have reasonable values for each parameter, note these as the initial "best guesses" to be used to run the cost function to get the optimal parameter values.

(c) Use `modFit` as in Listing 4.8 using the best guess parameter values obtained in the previous step. Report the values of a and b obtained from the optimization algorithm.

(d) Plot the data and overlay the model solution using the calibrated values of a and b by modifying Listing 4.9. Your model output should look like Fig. 4.1.

(e) Is this model a good description for this particular tumor growth pattern? Explain and indicate why an increasing value for K makes sense given these data. How could K increase with tumor size?

4.2.3 Sensitivity analysis

When we study larger systems of equations with more unknown or inaccessible variables and parameters, it is important to identify those factors that have the greatest effect on the model output, those that have similar effects, and those that can vary widely with little or no effect on the overall system. We also need ways to validate

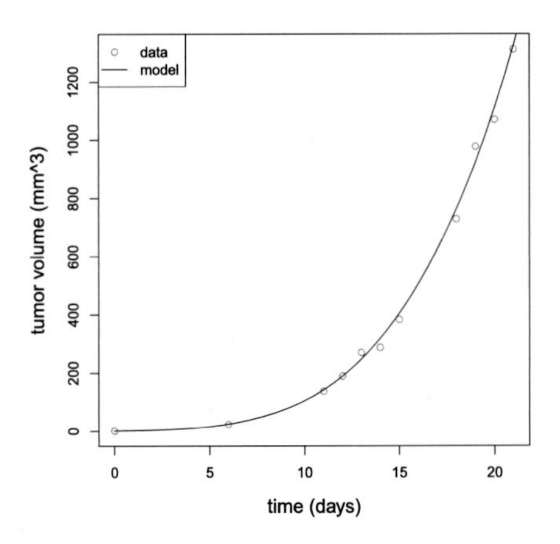

FIGURE 4.1

Data and model output using the logistic model with a dynamic carrying capacity in Exercise 4.2.

the model. Sensitivity analysis helps us accomplish these goals. For example, if we are using a model to produce some desirable output (system control), then sensitivity analysis can identify parameters that have the greatest effect on that output. Sensitivity analysis can also assist with research design; more effort should be devoted to obtaining estimates of the most sensitive parameters.

If sensitivity analysis reveals that the model output is extremely sensitive to small changes in parameters (i.e., below the noise in the data or the sensitivity of the instrumentation used), then we might question the validity of the model itself. Sensitivity analysis also reveals nonidentifiable parameters, which are parameters that have similar effects on the output and therefore are difficult to fit simultaneously. If resolving them is important to understanding the system, this mathematical conclusion points to more effort needed to attaining one or both of those parameter values experimentally.

Sensitivity analysis uses mathematical methods to study how changes in one or more parameters affects the model output, where model output could be one or more variables at a fixed time or over a continuous time, or some aspect of the model output such as the peak value of one variable. For example, recall that we might want to forecast when a population hits a certain value (whooping crane population recovery) or the trajectory of tumor growth over time to make a surgical decision or model the maximum concentration of a drug in the body given a dosing regime. The model output of concern should be aligned with the model objective, that is, the biological question being asked. Moreover, we may be interested in studying the behavior of the system over a large range of plausible parameter values, called *global sensitivity analysis*, or we may only be interested in how sensitive the output is to small perturbations in a parameter, which is referred to as *local sensitivity*. Next, we will cover examples of both global and local sensitivity analysis and the types of questions this analysis can help us answer.

Global sensitivity

If we are interested in how a model output variable is affected by changes in one parameter, we might first choose a set of values and plot the model output variable for each of these. A first simple exercise is to consider the logistic differential equation and how the population is affected by changes in the per capita growth rate and the carrying capacity. Then we can perform a similar exercise using the dynamic carrying capacity model.

Exercise 4.3

Again, we consider the logistic differential equation $y' = ry\left(1 - \frac{y}{K}\right)$.

(a) Sketch a solution $y(t)$ (by now, we should be able to sketch the general shape of the curve!). Then, on the same plot, sketch solution curves for increasing values of K. Confirm your understanding by solving the logistic differential equation in R (choose your own values for r, K, and $y(0)$) and plotting solution curves for different values of K on the same graph. Compare your numerical results to your sketch.

(b) Do the same exercise as part (a) but now vary r.

(c) Do the same exercise as part (a) but now vary $y(0)$.

In the previous sections, we considered the tumor growth model with a dynamic carrying capacity $\frac{dV}{dt} = aV\log\left(\frac{K}{V}\right)$, $\frac{dK}{dt} = bV^{2/3}$. The parameters a and b can vary from patient to patient, and these parameters can also be changed due to treatment strategies. We might be interested in how the tumor volume is affected by changes in these parameter values. In Listing 4.10, this system of differential equations is solved for a set of values for b, and the corresponding tumor volumes are plotted over time. Note that since we are repeatedly solving the differential equation, we can use a `for loop` to make our code efficient! The output of Listing 4.10 is shown in Fig. 4.2A.

Listing 4.10 (unit4ex10.R)

```
# Create vector with values of b used in simulations
b.vals = seq(0,2.5,by=.5)
n = length(b.vals)

# Set up empty plot
plot(NULL,xlim=c(0,20),ylim=c(0,1300),xlab="time (days)",ylab="tumor volume (mm^3)")

# Solve differential equations for each value of b
for (i in 1:n){
  parms = c(a=0.52,b=b.vals[i])
  out = tumorV(parms, times=seq(0,20), V0=1, K0=2.6)
  lines(out$time,out$V,type="l",col=i,lty=i)
}
legend("topleft",c("b=0","b=0.5","b=1","b=1.5","b=2","b=2.5"),lty=1:n,col=1:n)
```

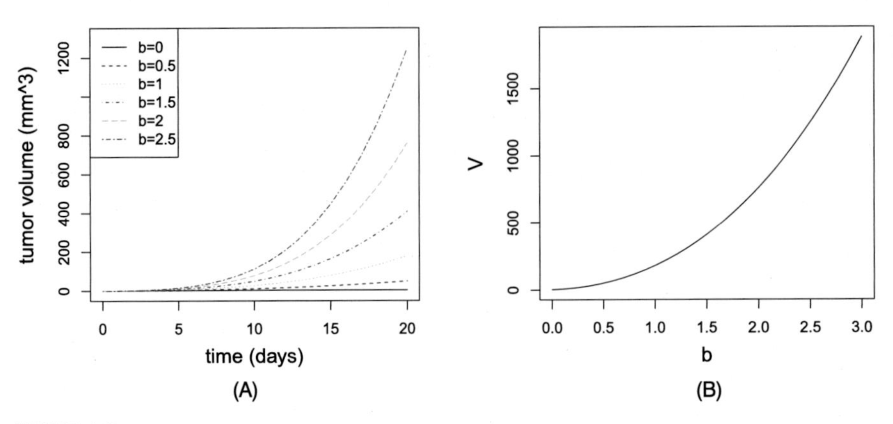

FIGURE 4.2

(A) Solutions to the dynamic carrying capacity tumor model for different values of b. (B) Tumor volume at day 20 as a function of b.

Exercise 4.4

Similar to Listing 4.10, simulate the tumor model with a dynamic carrying capacity but now vary a instead of b. Set $b = 2.37$.

In the two previous exercises, we were interested in the sensitivity of a model output variable over time, but sometimes we are interested in some aspect of the model output, such as a maximum value or the time at which a maximum occurs. Recall that in the garlic mustard case study in Unit 2, we plotted the long-term population as a function of the mortality rate parameter. This was an example of global sensitivity analysis since we considered the effect of a range of mortality rates on a model output, the long-term population size.

In the tumor model with a dynamic carrying capacity, we might be interested in plotting the tumor size on day 20 as a function of a model parameter. Similar to Listing 4.10, we can use a `for loop` to solve the system of differential equations for different values of b for $t = 0, \ldots, 20$. Then we can store the tumor volume V on day 20 in a vector, which allows us to plot the tumor volume versus b. Listing 4.11 produces the curve in Fig. 4.2B.

Listing 4.11 (unit4ex11.R)

```
b.vals = seq(0,3,by=.1)
n = length(b.vals)
tumor.size.20 = rep(0,n)  # Vector will store tumor size at day 20 for each value of b
for (i in 1:n){
  parms = c(a=0.52,b=b.vals[i])
  out = tumorV(parms, times=seq(0,20), V0=1, K0=2.6)
  tumor.size.20[i] = out$V[length(out$V)]
}
plot(b.vals,tumor.size.20,type="l",xlab="b",ylab="V")
```

Exercise 4.5

Following Listing 4.11 as an example, plot the tumor volume at day 20 as a function of the parameter a for values between 0 and 1. Discuss your results.

If we are interested in some aspect of the model output and we want to compare the effect of changing the values of different model parameters, then a nice way to visualize this is using a bar plot. For example, in Fig. 4.3, we compare the percent change in tumor volume at day 20 when the parameters are increased by 5%. Although we could just report these numbers, a bar plot can be a nice way to visualize these results when there are many parameters and you want to compare the effect of changing each parameter.

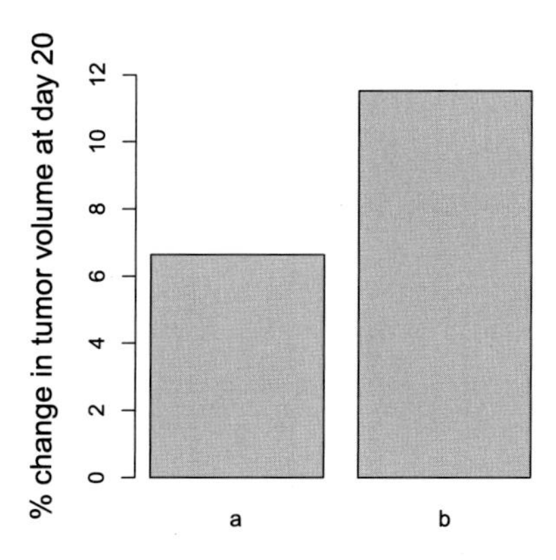

FIGURE 4.3

Percent change in tumor volume at day 20 when the parameter a or b is increased by 5%.

How do we create this bar plot? Since we are changing the values of the model parameters, the tumor volume is now considered to be a function of time and parameters a and b, that is, $V(t, a, b)$. If b increases by 5%, then the percent change in volume at day t is given by

$$\text{percent change in tumor volume} = 100\frac{V(t, a, b + 0.05b) - V(t, a, b)}{V(t, a, b)}, \qquad (4.4)$$

where we would use the original parameter values $a = 0.52$ and $b = 2.37$, and $t = 20$ for day 20. Then $V(20, 0.52, 2.37)$ is the volume at day 20 for the original parameter values, and $V(20, 0.52, 2.37 + 0.05 \cdot 2.37)$ is the volume at day 20 when b is increased by 5%. To calculate this in R, we need to solve the differential equation to compute the volume at day 20 using the original parameter values, and then we will need to solve the differential equation again using $b + 0.05b$. The process is repeated when increasing a by 5%. Later, we will create a bar plot for a model with many parameters! Therefore for practice, we will implement this sensitivity analysis using a `for loop`, as shown in Listing 4.12.

Listing 4.12 (unit4ex12.R)

```
parms = c(a=0.52,b=2.37) # original parameter values
out = tumorV(parms, times=seq(0,20), V0=1, K0=2.6) # solve using original parameter values
V.20 = out$V[length(out$V)]  # V(20) for original parameter values

Perc.change = rep(0,length(parms))  # set up vector to store percent change in V(20)

for (i in 1:length(parms)){
  perc = .05    # increase each parameter by 5%
  parms.pert = parms  # rename vector of parameters, don't change our original vector!
  parms.pert[[i]] = parms.pert[[i]] + perc*parms.pert[[i]]  # 5% change in parameter

  # solve system using new value of parameter
  out.pert = tumorV(parms.pert, times=seq(0,20), V0=1, K0=2.6)

  # select V at time 20 which corresponds to last entry of V
  V.pert = out.pert$V[length(out.pert$V)]

  # calculate percent change in V and store this in vector Perc.change
  Perc.change[i] = 100*(V.pert - V.20)/V.20
}

barplot(Perc.change,ylim=c(0,12),ylab="% change in tumor volume at day 20",
        names.arg=c("a", "b"))
```

Local sensitivity analysis

Local sensitivity analysis is a derivative-based approach to looking at sensitivity, and it is a key tool for exploring parameter identifiability, that is, whether or not it is possible to infer parameters from a given data set. In the previous example, we calculated the percent change in an output variable V due to some percent change in a parameter b. In general, consider some output variable $y(t)$ that depends on n parameters $p_1, \ldots, p_j, \ldots, p_n$. Similar to the previous example, consider a small change in parameter p_j and denote this change by Δp_j. The percent change in y is given by

$$\frac{y(t, p_j + \Delta p_j) - y(t, p_j)}{y(t, p_j)}. \tag{4.5}$$

This is similar to Eq. (4.4), but instead of including all of the parameters in the argument of y, that is. $y(t, p_1, \ldots, p_j + \Delta p_j, \ldots, p_n)$, only the parameter that is varied is shown. When comparing the effect of changing various parameters, it can be useful to normalize Eq. (4.5) by dividing by the percent change in the parameter value, that is,

$$\frac{\left(y(t, p_j + \Delta p_j) - y(t, p_j)\right) / y(t, p_j)}{\Delta p_j / p_j}, \tag{4.6}$$

which can be rewritten as

$$\left(\frac{p_j}{y(t, p_j)}\right)\left(\frac{y(t, p_j + \Delta p_j) - y(t, p_j)}{\Delta p_j}\right). \tag{4.7}$$

This expression gives the relative sensitivity. For example, if the value of this expression at a particular time is 2, then this means that a 1% increase in p_j results in about a 2% increase in y. If the value is -2, then a 1% increase in p_j results in a 2% decrease in y_i. Eq. (4.7) is an approximation to the *normalized sensitivity function*, which we call S_j, because some small perturbation Δp_j is chosen. If Δp_j is taken to be smaller and smaller, that is, the limit as Δp_j goes to zero, then we obtain the normalized sensitivity function given by

$$S_j(t) = \frac{p_j}{y(t, p_j)}\left(\lim_{\Delta p_j \to 0} \frac{y(t, p_j + \Delta p_j) - y(t, p_j)}{\Delta p_j}\right). \tag{4.8}$$

The expression inside the parentheses should look familiar from your calculus course! It looks like the definition of a derivative! However, since y depends on more than one variable (recall that it depends on t and parameters $p_1, \ldots, p_n$), a *partial derivative* is used, denoted by ∂ instead of d:

$$S_j(t) = \frac{p_j}{y}\frac{\partial y}{\partial p_j}. \tag{4.9}$$

Note that since y is a function t, the normalized sensitivity is also a function of t. If we have an explicit expression for y_i, then we can compute the partial derivative (to take a partial derivative with respect p_j, treat all other variables as constants) to obtain the sensitivity function. However, if only a differential equation for y_i is given and the explicit solution cannot be obtained, then the approximation in Eq. (4.7) may be used (referred to as the finite perturbation method). These normalized sensitivity functions can be computed for each parameter, and if we observe similar sensitivity functions for two parameters, then we should be cautious in fitting both parameters simultaneously; this analysis helps us determine if parameters are identifiable. We will explore this further at the end of this section. In this text, we will use a built-in function that computes the local sensitivity functions for us but refer to [78] for a more thorough introduction to solving the sensitivity functions.

Local sensitivity example: logistic model

In the previous section, the logistic equation was calibrated to a set of data, and the best-fit parameters were $r = 0.169$ and $K = 6.02$. A local sensitivity analysis can be performed to study the sensitivity of the model output $y(t)$ to the parameter values r and K. In other words, we want to compute the sensitivity functions

$$S_r(t) = \frac{r}{y}\frac{\partial y}{\partial r}, \qquad S_K(t) = \frac{K}{y}\frac{\partial y}{\partial K}. \tag{4.10}$$

The built-in function sensFun is part of the FME package, which uses the finite perturbation method (Eq. (4.7)) to approximate these sensitivity functions. Make sure you have installed and loaded this package before trying to use sensFun.

In Listing 4.13 the parameters are defined in the first line (these were found in the previous section). In the second line the function sensFun takes as input a function that returns a matrix or data frame with independent variable t in the first column and the output variables in the adjacent columns (in this example, y in the second column). This function has already been defined as logistic in Listing 4.3. Note that this previously defined function logistic has a vector of parameter values passed as the first argument, which is required when using sensFun. The second input must be the parameter values. Then the time vector and initial value are included as input since these are inputs into the function logistic.

Listing 4.13 (unit4ex13.R)

```
parms = c(r=0.169, K=6.02)
Sens.logistic = sensFun(logistic, parms, times=seq(0,80,by=0.1), y0=1.05)
head(Sens.logistic)
plot(Sens.logistic)
summary(Sens.logistic)
```

The third line in Listing 4.13 produces the following output:

```
> head(Sens.logistic)
    x var          r           K
1 0.0   y 0.00000000 0.000000000
2 0.1   y 0.01391110 0.002964159
3 0.2   y 0.02773857 0.005960873
4 0.3   y 0.04148105 0.008990240
5 0.4   y 0.05513700 0.012051957
6 0.5   y 0.06870529 0.015146674
```

We see that the output of sensFun is a data frame with the time steps in the first column, the variable name in the second column (this is helpful when there is more than one variable), the normalized sensitivity to parameter r in the third column, and the normalized sensitivity to parameter K in the last column.

The fourth line in Listing 4.13 plots the output of sensFun, that is, it plots the sensitivity functions as shown in Fig. 4.4. How do we interpret this information? First, note that the sensitivities are always positive since an increase in the intrinsic growth rate r or an increase in the carrying capacity K each results in an increase in the population y. The sensitivity to r goes to zero, because the population will approach the carrying capacity K no matter the value of r. Once the population is at its carrying capacity, r has no effect on the population size. At this time, we also observe that the sensitivity to K is one. Recall that the normalized sensitivity is interpreted as the percent change in the output variable over the percent change in a parameter. Therefore a sensitivity of one means that the population increases by one percent when the carrying capacity increases by one percent. In other words, when the population has reached the carrying capacity, a small change in the carrying capacity will result in the same change in population. One application of sensitivity analysis is to aid in experimental design. For this example, sensitivity analysis reveals that accurate estimates of both r and K from population data require reliable data in the early and midphases of growth and in population density values close to the maximum or K.

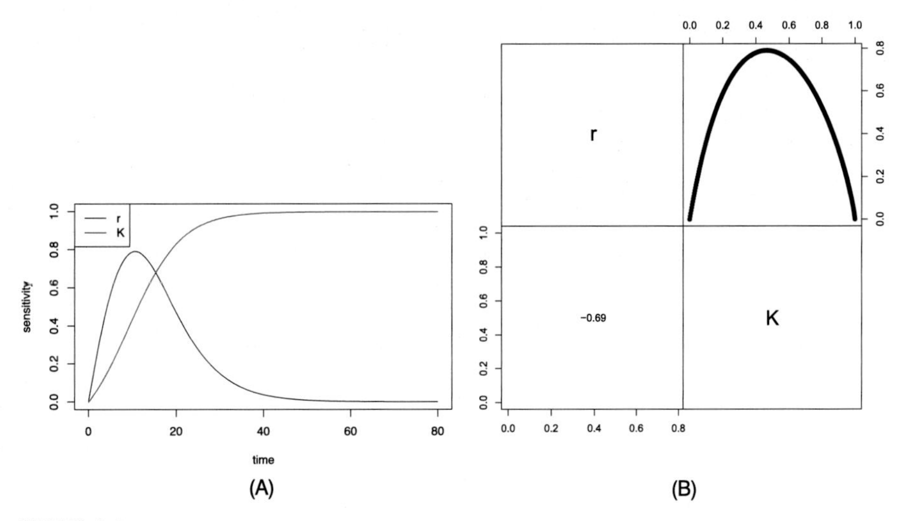

FIGURE 4.4

(A) Normalized sensitivity functions for the logistic differential equation. (B) Output demonstrating the extent to which the sensitivities of two parameters are correlated. The sensitivity of r is indicated on the y-axis in the top right panel and for K on the x-axis. The correlation coefficient between the two sensitivities is given in the lower left panel, and, as can be seen by visual inspection in this case, the correlation is poor, suggesting that r and K are identifiable parameters.

The sensitivities shown in Fig. 4.4 are functions of time, but it is often useful to calculate summaries of these sensitivities. The last line in Listing 4.13 produces the following summary statistics of the sensitivity functions:

```
> summary(Sens.logistic)
  value scale   L1     L2 Mean Min  Max   N
r  0.17  0.17 0.20 0.012 0.20   0 0.79 801
K  6.02  6.02 0.84 0.031 0.84   0 1.00 801
```

The column `value` gives the original values of the parameters used. The column `scale` is the value of each parameter used to normalize the sensitivity (see Eq. (4.10)); this will be the same as the value of the parameter unless specified otherwise as input to the `sensFun` function. The last four columns give the mean, minimum, and maximum of the sensitivity functions over time, and the number N of elements in the sensitivity output (which is equivalent to the number of time steps we defined). The columns L1 and L2 are two different measures of the overall sensitivity. The L1 result is just the mean sensitivity over time, whereas the L2 averages the square root of the sum of squared sensitivities over time. These summarizing measures provide a means to rank the importance of various parameters. In this example the results show that the model is much more sensitive to K than to r. Larger values of L1 and L2 for a given parameter correspond to a greater sensitivity or response in y to a perturbation in that parameter.

To determine if parameters r and K have a similar sensitivity on the model output, their sensitivity functions shown in Fig. 4.4 can be compared graphically, and their correlation can be estimated using the command `pairs`:

```
pairs(Sens.logistic)
```

which produces the result shown in Fig. 4.4B. The curve in the upper right corner is plotting the pairwise relationship between the curves in Fig. 4.4, where the vertical axis is the sensitivity of V to parameter r, and the horizontal axis the sensitivity of V to parameter K. Here $t = 0$ corresponds to the point (0,0), and $t = 80$ corresponds to the point (1,0) on this curve. If this curve was close to linear, it would mean that the sensitivities to these parameters are highly correlated. The strength and direction of this linear relationship are measured using the correlation coefficient given in the lower left box. The correlation coefficient ranges from -1 to 1, where a value close to one means that there is a strong positive linear correlation, and a value close to negative one means that there is a strong negative linear correlation. High correlations indicate that the corresponding parameters have equivalent effects on the output variable. This means that it is questionable to jointly estimate these parameters using data on that particular variable! We say that these parameters might be *nonidentifiable parameters*, which means that they cannot be determined with sufficient precision using only data for the output variable. This can be common in complex models and sometimes unavoidable! If the square of the correlation coefficient is greater than 0.85, then we usually say that these parameters are highly correlated, which is clearly not true in the example for r and K.

Exercise 4.6

Perform a local sensitivity analysis on the tumor model with a dynamic carrying capacity. Use $a = 0.52$, $b = 2.37$, $V(0) = 1$, and $K(0) = 2.6$ and consider the times $t = 0, \ldots, 20$.

(a) Use Listing 4.13 as a template and set `Sens.tumor = sensFun(...)`. Type `head(Sens.tumor)` and `tail(Sens.tumor)` and explain what each column represents.
(b) The sensitivity function computes sensitivities to both $V(t)$ and $K(t)$. Plot only the result for $V(t)$ by typing `plot(Sens.tumor,which="V")`. Interpret your results.
(c) Look at the summary measures. What are your conclusions based on the summary measures?
(d) Plot the sensitivity to a versus the sensitivity to b by using the `pairs` command. Again, use `which="V"` to look at only the sensitivities of the parameters to the tumor volume. What can you conclude from your result?

In this section, we introduced multiple ways to assess the sensitivity of a model to one or more parameters. One rarely uses all approaches with a single model! Instead, we choose global or local, depending on the question at hand. We plot the data to help us understand what parameters affect the output of interest, which itself may vary depending on whether you are deciding on a strategy, say for cancer treatment, or trying to disentangle some of the steps of the process, or trying to assess how confident you are about the parameter estimates. In the example and cases that follow, we will practice choosing and applying different strategies.

4.2.4 Putting it all together: the dynamics of Ebola virus infecting cells

Here we will implement various tools introduced in this chapter to numerically solve a system of differential equations, estimate unknown parameter values using an optimization package, and perform a sensitivity analysis to understand how each parameter affects the model output. Our system is of great interest and urgency worldwide, the process by which a virus infects host cells that respond by generating a new virus. These steps are fundamental to understanding how viruses propagate and spread throughout a population. Although there are multiple emerging diseases to explore, we will focus on Ebola, which frequently causes death due to its ability to infect its host cells and propagate. Our goal is to develop and calibrate a model to help us understand the system, and then explore the effect of various parameters on the virus population as a means to better understand treatment strategies.

First, we must define our model. Here we are considering experiments in which cells growing in the lab are inoculated with a known small amount of virus as a model for virus infecting the cells in the human body. Once the Ebola virus infects a cell, it uses the cellular machinery to make many new copies of itself that emerge when the cell bursts. Thus we have to follow the dynamics of uninfected cells, infected cells, and the virus.

This system can be described by the following three equations [46]:

$$\frac{dU}{dt} = \lambda - \alpha U - \beta U V, \tag{4.11}$$

$$\frac{dI}{dt} = \beta U V - \delta I, \tag{4.12}$$

$$\frac{dV}{dt} = pI - cV, \tag{4.13}$$

where $U(t)$ is the number of uninfected susceptible cells per mL, $I(t)$ is the number of infected cells per mL, and $V(t)$ is the concentration of virus measured in foci forming units per mL (ffu/mL). The variable $V(t)$ is easily measured (similar to measuring colony-forming units to count bacteria in the lab in Unit 2). Virus infects susceptible cells at rate β, infected cells are cleared at rate δ, infected cells release virus at rate p, and virus particles are cleared at rate c. The susceptible cells are produced at a constant rate λ and die at rate α.

Exercise 4.7

(a) Sketch a diagram representing the three variables needed to describe cell-level infection and the relationships among them using the sketch in Fig. 4.5 as a prompt.

(b) Does the model in Eqs. (4.11)–(4.13) make sense given the sketch you made of the system? Make a table of units for each of the parameters and check your answers with Table 4.1.

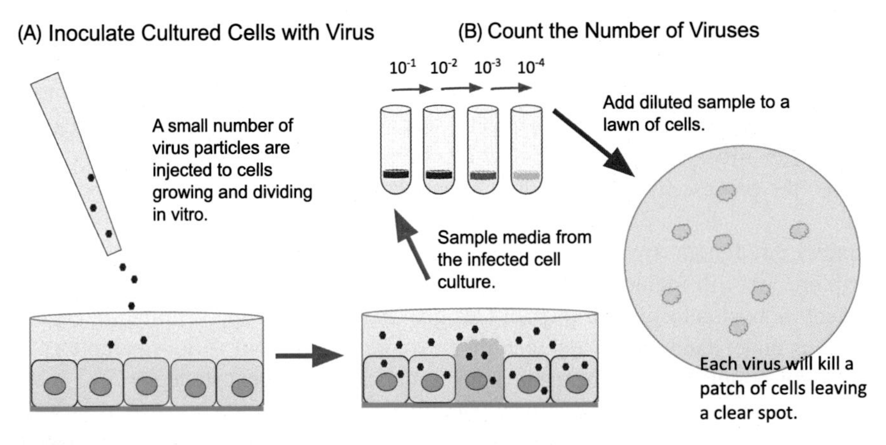

FIGURE 4.5

(A) When virus is added to uninfected cells growing on a plate, some cells become infected allowing the virus to reproduce and then emerge by disrupting the cells. (B) The number of viral particles that are generated can be counted by sampling the culture medium, diluting it significantly by serial dilution and then adding the diluted media to a lawn of uninfected cells. Here there were five viral particles in the dilution creating seven clear spots on the culture dish due to cell death. This is equal to 7×10^4 particles in the original sample.

This model in Eqs. (4.11)–(4.13) is a useful means of predicting how the viral load, that is, amount of virus in our culture (or in an infected individual), will vary in response to clinical interventions. Many clinical researchers are seeking interventions to minimize the amount of new virus formed, maximize its clearance from the body, and minimize the number of infected cells. Therefore, as you explore the model, see if you can identify ways to minimize both the virus and infected cell numbers.

Numerical solution for the Ebola model

Now that the model is formulated, we want to make sure we can solve the system numerically. The parameter values in Table 4.1 can be used to do this, but we will fit some of these parameters later based on data.

Exercise 4.8

(a) Solve Eqs. (4.11)–(4.13) in R using the parameter values in Table 4.1. Refer to Listing 4.4 to set up these commands in R. Initial conditions need to be defined. It is reasonable to begin with the following values for the variables: $U(0) = 5 \times 10^5$ cells per mL, $I(0) = 0$ cells per mL, $V(0) = 9$ foci forming units per mL. Look at the first six rows of the output using the `head` command. What does each column represent?

(b) After solving the system, plot the uninfected cells $U(t)$, infected cells $I(t)$, and virus particles $V(t)$ over time. Compare your result to Fig. 4.6.

Table 4.1 Initial guesses for parameter values for the Ebola virus infection model.

Parameter	Definition	Guessed value
λ	uninfected cell growth rate constant	500 cells day^{-1}
α	natural death rate constant of uninfected cells	0.001 day^{-1}
β	viral infection rate constant	10^{-7} (day · cell)$^{-1}$
δ	natural death rate constant of infected cells	0.001 day^{-1}
p	growth rate constant for the virus in infected cells	500 day^{-1}
c	natural viral clearance rate constant	5 day^{-1}

Fitting parameters to the Ebola model: calibrating the model

The parameter values given in Table 4.1 are based on other studies with other cell types and viruses. They are good guesses, but our job is to find better estimates for this particular experimental system. To do this, we need data. The most accessible data are the measurements of the number of viral particles over time. Fortunately, we have a reasonable data set (simulated but realistic) taken at daily intervals over 11 days (from day 0 through day 10), which we can enter as

```
V_data = c(9, 5169, 213355, 1248299, 722206, 11729657,
           21914277, 23494465, 22858724, 22403262, 23357682)
```

Now we need to make some assumptions. First, we will use the same initial values as before, that is, $U(0) = 5 \times 10^5$ cells per mL, $I(0) = 0$ cell per mL, and $V(0) = 9$ ffu/mL, which are reasonable for typical cell cultures and the dose of virus that was used to inoculate the system. We assume that the literature values for two of our parameters are reasonable. The parameter α is given in literature as 0.001 day^{-1} [44], and the natural infected cell death rate constant δ is assumed to be 0.001 day^{-1} based on the half-life of epithelial cells in lungs [57]. If we assume that the system is at steady state, then we can also calculate a value for λ. At steady state, $dU/dt = 0$, $dI/dt = 0$, and $dV/dt = 0$, and if there is no virus, then $V = 0$ and $I = 0$, which requires $\lambda = \alpha U(0)$. Using our values for α and $U(0)$, we have $\lambda = 500$ cells per day.

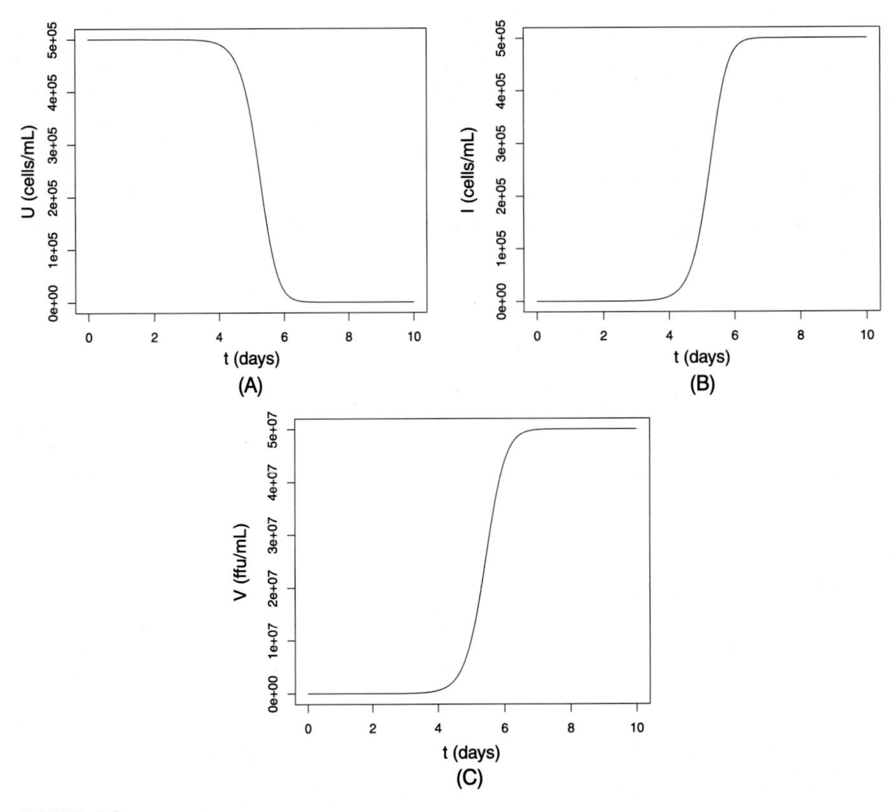

FIGURE 4.6

Ebola model output where (A) $U(t)$ is the number of uninfected suseptible cells per mL, (B) $I(t)$ is the number of infected cells per mL, and (C) $V(t)$ is the concentration of virus measured in foci forming units per mL (ffu/mL).

The other parameters, β, p, and c are more difficult to obtain experimentally, and therefore we must use our cost and optimization functions to find best estimates for these three values. Refer to Listing 4.6 to set up the cost function. Recall that this function sets up a vector of squared residuals between the model output $V(t)$ for a given set of parameters and the actual viral data, the only variable we can measure. Then we used `modfit` to find the set of parameter values that minimize this cost function. However, `modFit` can only do its job effectively if it has a reasonable set of guesses for the unknown parameters. To obtain reasonable guesses, we can use `manipulate` with sliders for each of the unknown parameters as in Listing 4.7.

Exercise 4.9

Use the manipulate function to create sliders for β, p, and c, and inside the manipulate function, plot the model output $V(t)$ along with the data set. Reasonable ranges for the sliders are 0 to 10^{-6}

stepping by 10^{-7} for β, 0 to 1000 for p, and 0 to 20 for c. After playing with the sliders, write down your best guess for each parameter value, adding them to Table 4.1.

Now we can calibrate the model to attain estimates for these three parameter values that minimize the squared residuals between the model output and data.

Exercise 4.10

(a) Write a cost function and run `modfit` with your best guesses from the previous exercise. To set this up, refer to Listings 4.6 and 4.8. What are your estimates for β, p, and c? Add these new values labeling them "fitted parameter values" to your table of estimates. How different are these from your guesses attained using the manipulate function?

(b) Plot the data and overlay the solution to the calibrated model. Refer to Listing 4.9 to set up the code to do this, and check your result with Fig. 4.7.

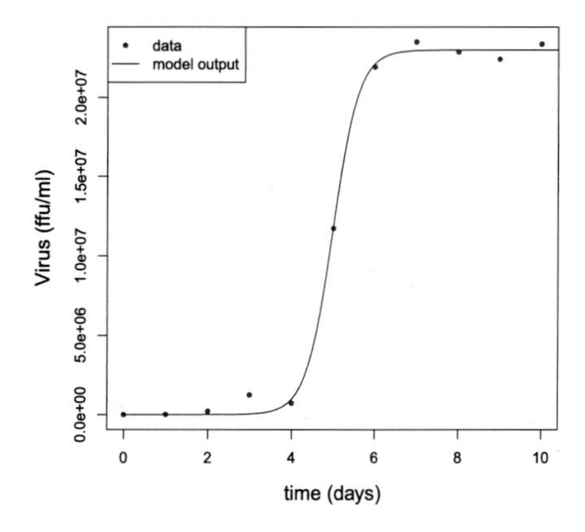

FIGURE 4.7

Ebola model output compared with the concentration of virus over time.

Local sensitivity analysis: assessing key parameters in the Ebola virus-cell system

Recall that one of the key goals of modeling the virus-cell infection dynamics is minimizing virus production rates and damage to the cells. We know that our model has multiple parameters and exploring the relative sensitivity to each of them can help guide decisions about the next stages of clinical research or reasonable treatment strategies. We also know that sensitivity analysis is a good check on our model since extreme sensitivity to small changes in one or more parameters suggests something could be wrong.

We can approach the sensitivity analysis as we did for the logistic model. However, the system of equations modeling Ebola virus infection has more than one variable. Therefore we can calculate the sensitivity of each model output variable to each model parameter. The three variables are $U(t)$, $I(t)$, and $V(t)$, and the six model parameters are λ, α, β, δ, p, and c. We could look at the sensitivity functions for uninfected and infected cells, but since our goal is minimizing the number of virus particles formed, we will look at the sensitivity of only $V(t)$ due to small changes in the parameters. A second reasonable goal might be minimizing the number of infected cells $I(t)$ because these are the direct cause of tissue damage in this disease.

Exercise 4.11

(a) Use the built-in function `sensFun` to compute the local sensitivity functions. Refer to Listing 4.13 and Exercise 4.6 for examples on how to this. Plot the sensitivity functions for only V (use `which="V"` to choose only V) (see Fig. 4.8).

(b) Look at the summary of the sensitivity results for each parameter over the time course of the infection. What conclusions can you make by comparing the summarizing sensitivities?

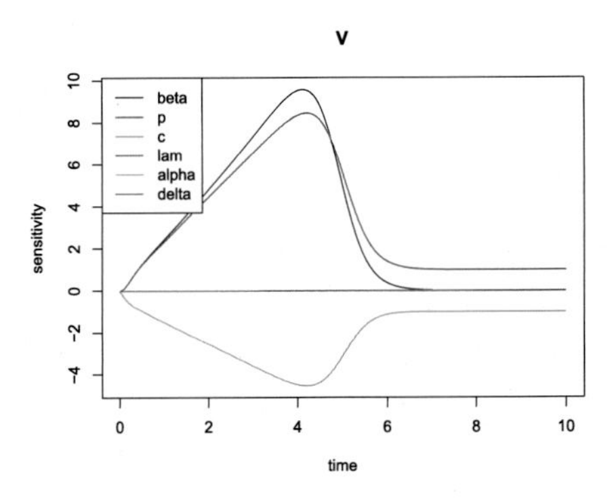

FIGURE 4.8

Sensitivity of model output $V(t)$ to model parameters.

The sensitivity functions shown in Fig. 4.8 show positive values for β and p, which means that small increases in the infection rate and the virus release rate from infected cells increase $V(t)$. The sensitivity to c is negative, which implies that a small increase in the virus particle clearance rate decreases V. The sensitivity to small changes in λ, α, or δ is almost zero. This means that these parameters have little effect on the overall model dynamics, and therefore it is adequate to fix these parameters at the given values. However, it is clear from the sensitivity functions and the summary results that β, p, and c govern the dynamics of this system. Fig. 4.8 also reveals that β and p seem to have a similar effect on the model output, and c has a large but negative effect that mirrors that of β and p. These pairwise relationships can be visualized using the `pairs` command as shown in Fig. 4.9.

Here the sensitivity functions for $V(t)$ are used in Fig. 4.9, but sensitivity functions for all of the variables or combinations of them can also be used. To interpret the graphical results, look at the second box from the left in the top row in Fig. 4.9. The

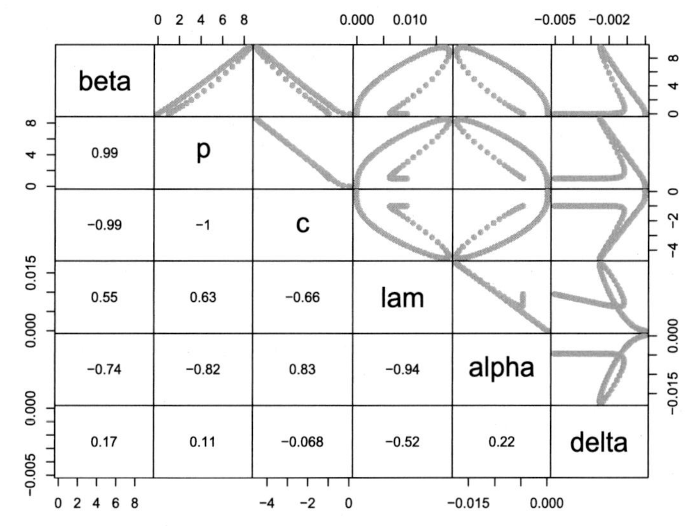

FIGURE 4.9

Pairwise plot of sensitivity functions for $V(t)$ to the right of the diagonal along with the correlation coefficients listed in rows to the left of the diagonal.

sensitivity to β over time is plotted against the sensitivity to p over time. Note that the relationship is almost linear! The strength and direction of this linear relationship is measured using the correlation coefficient r, given in the first box in the second row. Recall that the correlation coefficient is between -1 and 1, where r close to one means that there is a strong positive linear correlation. Note that the correlation coefficient for β and p are very close to 1, implying that these parameters are highly correlated, and the correlation coefficient for p and c is close to -1, which means that there is a strong negative linear correlation. The sensitivity functions for β, p, and c are strongly correlated ($r^2 > 0.85$), and the sensitivity functions for λ and α are also highly correlated. High correlations indicate that the corresponding parameters have similar effects on the output variable. For example, increasing β by some small percentage will have an equivalent effect to increasing p or decreasing c by that same percentage. This means that it is questionable to jointly estimate these parameters only using the virus data. We say that these parameter might be *nonidentifiable parameters*, which means that they cannot be determined with sufficient precision using only data for V.

Exercise 4.12

(a) Use the `pairs` command to produce pairwise plots of the sensitivity function as shown in Fig. 4.9.

(b) Based on this output, what would you recommend as the next most important parameter to try to measure directly? Why?

We now have an expanded modeling toolkit to help us understand and analyze some complex biological systems. In the following case studies, we work through the modeling process with these new tools to develop strategies to avoid future pandemics, test treatment strategies for prostate cancer, and understand how bioluminescence occurs in bobtail squid.

4.3 Case study 1: Modeling the 2009 influenza pandemic

Goals: We will become familiar with the classic epidemic model of an infectious disease in a large population. This SIR model divides a population into three dynamic classes: susceptible class S, infected class I, and removed or recovered class R. We will explore the consequences of varying key parameters, including the recovery rate and transmission rate, and calculate a useful value, called the effective reproduction number R_e, which measures the transmissibility of an infection and helps monitor the effectiveness of control strategies such as vaccination programs. Finally, we will use data from the 2009 influenza pandemic to estimate parameters and test public health interventions, which could be used to respond to an ongoing epidemic.

4.3.1 Background

Mathematical approaches helped inaugurate our modern approaches to combating disease. Florence Nightingale revolutionized battlefield medicine during the Crimean War (1854) and general nursing care by emphasizing the importance of collecting accurate data on the number of ill individuals, their diseases, and mortality rates. She revolutionized presenting data in ways that could be readily understood. Synchronously, Dr. John Snow responded to the 1854 cholera outbreak in London with his now famous map of individuals with cholera that traced the source of the disease back to a single water pump. He famously forecasted: "*the time will arrive when great outbreaks of cholera will be things of the past; and it is the* ***knowledge of the way in which the disease is propagated*** *which will cause them to disappear*" [28].

Epidemics are defined as a rise in the number of people afflicted with an infectious disease, and epidemiologists are the scientists who prepare for, document, and respond to an outbreak. Florence Nightingale and John Snow did their work before it was accepted that infectious agents (viruses, bacteria) can transmit diseases from an infected individual to an uninfected individual directly or indirectly. Their reports helped lay the groundwork for our current knowledge of infectious disease. Specific public health measures depend on the biology of the disease agent, for example, on how it is spread or whether there are drugs to limit disease progression or vaccination opportunities. However, the fundamentals, systematic data collection, and communication are critical first steps.

What to do in any given epidemic depends on the local circumstances and the characteristics of infection. It is important that decisions be made quickly to stop the epidemic but should be in proportion to the real risk. We have seen multiple examples of rapid responses to emergent threats such as the Zika virus, which can cause brain underdevelopment if pregnant women are infected, Ebola (the virus we examined in the background to this chapter) with its deadly symptoms, as well as dengue and West Nile viruses. How do the epidemiologists figure out what to do? Often they begin with a version of the SIR model developed in 1927 by Kermack and McKendrik, who, like John Snow and Florence Nightingale, had the vision to suggest that understanding the underlying relationships quantitatively was essential to successfully limiting an epidemic:

In the course of time the epidemic may come to an end. One of the most important problems in epidemiology is to ascertain whether this termination occurs only when no susceptible individuals are left, or whether the interplay of the various factors of infectivity, recovery, and mortality may result in termination, whilst many susceptible individuals are still present in the unaffected population.

—*A Contribution to the Mathematical Theory of Epidemics*

To limit the number of individuals affected by a disease outbreak, it is necessary to predict if it is likely to reach epidemic proportions, how many people in the population will fall ill, and if or when the disease will die out. The first step is collecting data on the number of individuals in a given area who are ill as a function of time and then projecting the time course of the disease knowing the total population size and something about how the disease is spread. This is what was done in the spring of 2009 when a very unusual spring uptick in influenza cases occurred in Mexico. This "new" influenza, now called A(H1N1)pdm2009, spread rapidly from Mexico across the globe largely by air travel. By mid June, the World Health Organization declared a pandemic, and vaccine preparation began so it would be available by late October. Unlike normal seasonal flu, this flu appeared in summer. No one was protected by the annual flu shot, and younger people were more likely to be hospitalized or die than older people. Public health workers began action immediately, continuously collecting data, updating their models, and making decisions about how to limit the spread using a version of the SIR model.

4.3.2 The SIR model

The original purpose of the SIR model was to predict the number of individuals involved and the time course of an epidemic. When a virus like the influenza virus enters a new population, the dynamics of the disease can initially be understood by dividing the population into three groups, susceptible S, infectious I, and recovered R with the assumption that the recovered population has also gained immunity and will not get sick again in the time being considered. Individuals can move progressively through each of these compartments as suggested by the following diagram:

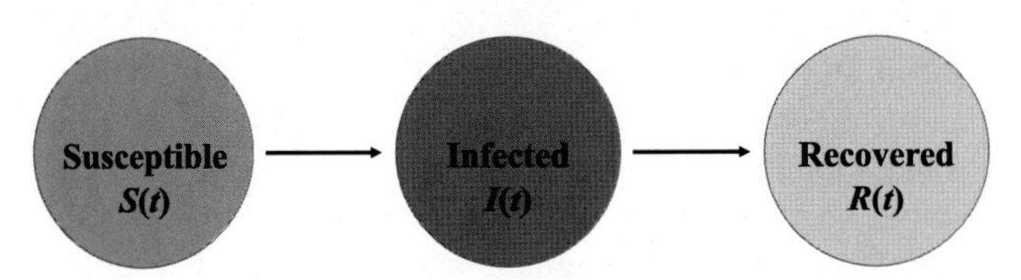

The flow rates between populations depend on how virulent or infectious the virus is and the duration of infection. To formulate a mathematical model, we need to understand how these three populations, the variables, interact. For example, how do infected individuals control the rate at which the susceptible people become ill? Which variables affect the rate of infected individuals moving to the recovered class?

The qualitative SIR model can be written as a system of differential equations

$$\frac{dS}{dt} = -bSI, \tag{4.14}$$

$$\frac{dI}{dt} = bSI - kI, \tag{4.15}$$

$$\frac{dR}{dt} = kI, \tag{4.16}$$

where b is the disease transmission rate, which incorporates the number of contacts an individual has per unit time, and k is the recovery rate. Note that k has units of 1/time, and therefore $1/k$ is the duration of infection. (Note the similarity between this model, which models how a viral infection travels through a population, and Eqs. (4.11)–(4.13), which examine the interaction between a virus and its host cells in a single individual. Compare and contrast these two models and the purpose of each model.)

Exercise 4.13

Understanding Model Assumptions

(a) What are the units of the expression bS? What is the interpretation of this expression and its assumptions? Note that mass action is used to model the transmission term bSI. What is the similarity between this and the Type I Holling response used in the predator–prey model, as discussed in Unit 3?

(b) Adding the three equations, we see that $\frac{dS}{dt} + \frac{dI}{dt} + \frac{dR}{dt} = 0$. What does this imply?

(c) The SIR model assumes a closed population and no natural births or deaths. Under what circumstances is this a reasonable assumption for this model?

Exercise 4.14

Modifications to the SIR Model

(a) Many infectious diseases have a latent period, which is the period between infection and when the individual becomes infectious. Add an "Exposed $E(t)$" compartment in the qualitative diagram to represent the latent period and write a corresponding set of differential equations. This is referred to as an SEIR Model.

(b) Some infectious diseases do not elicit immunity in response to an infection. In this case, there is no recovered class distinct from the susceptible class. Draw the flow diagram and the corresponding differential equations of the SIS Model.

(c) Vector-borne diseases are transmitted through an external vector which harbors an infectious agent; for example, malaria is transmitted to humans when mosquitoes carrying the malaria parasite bite. Sketch a qualitative diagram that shows the movement and interactions among compartments for humans and the vector. Assume that humans and vectors infect each other across populations, but not within populations, and that the vector host remain infectious while humans eventually recover.

Model simulations

Consider a typical community that has a flu outbreak over the winter. What happens to the susceptible population over time? If we consider the number of infected individuals over time, then what would this curve look like? In this section, we will use a given set of parameter values to explore the model output and whether or not the output matches our intuition. Moreover, we will try to understand how each parameter affects the course of an outbreak.

Exercise 4.15

(a) Since the population is constant, denoted by N, the number of recovered individuals can be calculated using $R(t) = N - S(t) - I(t)$, and the SIR model can be reduced to a system of two differential equations (4.14) and (4.15). Solve these two equations using the SIR model template given in Listing 4.14. Plot $S(t)$, $I(t)$, and $R(t)$ on the same plot.

(b) Discuss the shapes of the curves. Does the model output seem to capture the dynamics of a flu outbreak? Do the parameter values seem reasonable?

(c) How many individuals were infected during this outbreak? (*Hint*: All the infected individuals enter the recovered class!)

(d) How would you expect the graphs of $S(t)$, $I(t)$, and $R(t)$ to change if you increase the infection rate b and keep the recovery rate k fixed? Discuss this in terms of the interpretation of b.

(e) How would you expect the graphs of $S(t)$, $I(t)$, and $R(t)$ to change if you increase the parameter k and keep b fixed? Discuss this in terms of the physical interpretation of k.

(f) Confirm your predictions in part (d). Choose at least three different values of b and plot the respective outputs $I(t)$ on the same plot. Can you find a value of b that causes the infection to die out instead of resulting in an outbreak?

(g) Now confirm your predictions in part (e). Choose at least three different values of k and again plot $I(t)$ for each value on the same plot. Can you choose a value of k that causes the infection to die out?

Listing 4.14 (unit4ex14.R)

```
SIR = function(parms, times, S0, I0){
  with(as.list(c(parms)),{
    SIR.model= function(t,y,parms){
      with(as.list(c(parms,y)),{
        INSERT CODE HERE
      })}
    y0=c(S=S0, I=I0)
    out=ode(y0,times,SIR.model,parms)
    as.data.frame(out)
  })
}

S0=160000; I0=10; N=S0+I0  # Initial number of individuals in each class
times=seq(0,50,by=0.1)     # Time in units of days
parms=c(b=5e-6, k=1/3)     # Transmission rate (1/(day*individuals)), recovery rate (1/day)
SIR.mod=SIR(parms, times, S0, I0)

PLOT S(t), I(t), R(t)
```

4.3.3 Cumulative number of cases

To compare our model to real data, we need to consider what data are available. In the SIR model, $I(t)$ is the number of individuals infected at a given time t. However, data are typically given as number of new cases reported each day or each week. How can we compare our model output to the rate of new cases reported? If the data are new cases reported each day, then summing the rate of new cases from time 0 to the given time point will give the cumulative number of cases, which is easy to calculate in our model! The rate of individuals moving to the infected class is given by bSI. If we create a new variable $C(t)$ representing the cumulative number of infections at time t, then $C'(t) = bSI$. This differential equation can be added to the original SIR model to solve for the cumulative number of cases.

Exercise 4.16

Using the parameter values in Listing 4.14, modify your code to include the differential equation $\frac{dC}{dt} = bSI$ in the function `SIR`. Plot $C(t)$. Report the total number of individuals infected throughout the outbreak and the fraction of the population infected.

Now that we can solve for $C(t)$, it is easy to calculate the total number of individuals who have been infected up to some time. In particular, the value of $C(t)$ at a time following the end of an outbreak will give the total number of individuals infected during the outbreak. The "attack rate," which is the proportion of the population infected throughout an epidemic, is a commonly reported value. In terms of the model variables, the attack rate is given by C_∞/N where C_∞ represents the long-term value of C. It is interesting to see how this value depends on combinations of the model parameters b and k.

We can simulate the model over many combinations of b and k, then use a heat map to plot the fraction of the population who were infected during a given time period. Read carefully through Listing 4.15 to see how a nested for loop is used to simulate the model for various combinations of the parameters. The fraction of people infected as a function of b and k can be viewed as a heat map, shown in Fig. 4.10. (Recall that a nested for loop was used in the tumor growth case study in Unit 3 to plot the sum of the squared residuals for combinations of two parameters.)

Listing 4.15 (unit4ex15.R)

```
k.vals = seq(0.1,1,length.out=20)
b.vals = seq(1e-6,1e-5,length.out=20)
Frac.Inf=matrix(0,length(b.vals),length(k.vals)) # Set up matrix to store output
times=seq(0,100,by=0.1) # Simulate out to t=100 days
n=length(times)
S0=160000; I0=10; C0=10  # Initial conditions
N = S0+I0  # Total population

for (j in 1:length(k.vals)){
  for (i in 1:length(b.vals)){
    parms=c(b=b.vals[i], k=k.vals[j])
```

```
      SIR.mod=SIR(parms, times, S0=S0, I0=I0, C0=C0)

      # Choose C at last time point, divide by N to get fraction of population
      Frac.Inf[i,j]=SIR.mod$C[n]/N
    }
}

x = list(title = "recovery rate constant (k)") # x-axis label
y = list(title = "transmission rate constant (b)") # y-axis label

plot_ly(x=k.vals,y=b.vals,z=Frac.Inf,type="heatmap") %>%
  layout(title = "Fraction of individuals infected during outbreak", xaxis = x,
         yaxis = y)
```

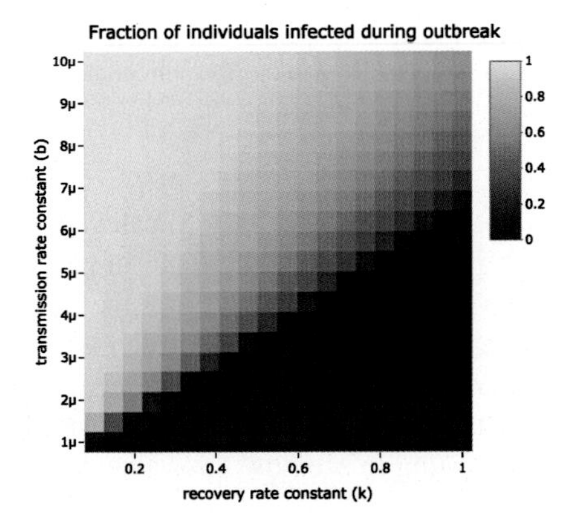

FIGURE 4.10

Fraction of individuals who have been infected after 100 days as a function of transmission rate and recovery rate.

Exercise 4.17

Interpret the results shown in Fig. 4.10. Is there some threshold for an epidemic to occur? Estimate this threshold in terms of the relationship between b and k that represents when an epidemic occurs.

4.3.4 Epidemic threshold

When an infected individual is introduced to a population of susceptible individuals, what conditions lead to an epidemic? Fig. 4.10 shows a possible relationship between b and k that gives a threshold necessary to result in an epidemic. Intuitively, we might say that an epidemic occurs if the infected individual infects at least one other individual during their time of infection. Mathematically, this is equivalent to saying that an epidemic occurs if $I'(0) > 0$. In the next exercise, we will start with this definition to derive the *basic reproduction number* R_0, the threshold that determines whether an infectious disease will quickly die out or cause an epidemic. This number is frequently reported in the literature.

Exercise 4.18

Here we will derive the basic reproduction number R_0 assuming that at time $t = 0$, there is one infected individual and the rest of the population is susceptible.

(a) At time $t = 0$ the rate of change of infected individuals is given by

$$I'(0) = bS(0)I(0) - kI(0). \tag{4.17}$$

If there is one infected individual in a population N, then $S(0) = N - 1$ and $I(0) = 1$. For a large population, $S(0) \approx N$. Replacing $S(0)$ with N in Eq. (4.17), we obtain

$$I'(0) \approx bNI(0) - kI(0). \tag{4.18}$$

If $I'(0) > 0$, then the infection invades the population and an epidemic occurs. If $I'(0) < 0$, then the infection does not spread to other individuals, and no epidemic occurs. Using Eq. (4.18), show that the threshold for an epidemic to occur is $\frac{bN}{k} > 1$. The expression on the left side of the inequality is denoted by R_0 and is called the *basic reproduction number*.

(b) The basic reproduction number $R_0 = \frac{bN}{k}$ can be understood by considering the interpretations and units of bN and $\frac{1}{k}$. Do this and explain why R_0 represents the number of secondary infections generated by a typical infected individual in an entirely susceptible population.

To summarize, the basic reproduction number $R_0 = \frac{bN}{k}$ is defined at the beginning of an epidemic when the entire population is assumed to be susceptible, and it gives a threshold that determines whether or not an epidemic occurs. If $R_0 < 1$, then $I(t)$ decreases monotonically to zero in time; if $R_0 > 1$, then an infection can invade the population and cause an epidemic. Notice how these statements relate to your interpretation of the heat map in Fig. 4.10.

However, in many cases, the entire population may not be susceptible, or we are in the middle of an epidemic, and we want to calculate the reproduction number at that point in time. A related value, the *effective reproduction number* $R_e(t)$ is the average number of secondary cases per primary case observed in a population with an infectious disease at any point in time. The effective reproduction number can be derived in a similar way as was done in Exercise 4.18, but we can no longer assume that $t = 0$ and $S \approx N$. Instead, we use the current susceptible population and find that

$$R_e(t) = \frac{bS(t)}{k}. \tag{4.19}$$

Exercise 4.19

Using the parameter values and initial conditions in Listing 4.14, solve the SIR model as you did before, but now use the results to plot the effective reproduction number versus time. Discuss the shape of the curve, the range of values for R_e, and how these are related to what is happening with the infected population $I(t)$.

4.3.5 Public health interventions

Understanding the effective reproduction number can help public health experts to predict strategies to prevent an epidemic or reduce the number of people becoming infected during an outbreak. For example, for influenza, the following strategies are often implemented:

- Flu vaccines;
- Public health campaigns to encourage frequent hand washing and face masks;
- Antivirals to reduce the duration of infection;
- Self-isolation, such as encouraging individuals to stay home from work or school.

Exercise 4.20

(a) Recall from the previous section that if $R_e(t)$ is less than one, then an epidemic does not occur. From Eq. (4.19), identify which parameters, if decreased or increased, might result in reducing R_e to below one.

(b) How would each strategy listed earlier (vaccines, hand washing, antivirals, self-isolation) affect the parameters b and k or the variable $S(t)$?

Public health measures are hard to implement because they depend on supplies and people. For example, a focused vaccination campaign may provide vaccine to a community over a very short time, or individuals might be able to get vaccinated at any time throughout an outbreak at a rate dependent on clinic capacity. We will simulate both of these strategies in the next exercises.

Exercise 4.21

(a) Assume that 30% of the population has been vaccinated by $t = 0$. This will decrease the initial number of susceptible individuals. Use the parameter values in Listing 4.14 to solve the SIR model and plot $I(t)$. Overlay $I(t)$ for the scenario when no one is vaccinated. How do they differ?

(b) Calculate the total number of individuals infected throughout the outbreak with and without a 30% vaccination rate. Compare the outcomes.

(c) Based on your model, what percentage of the population would need to be vaccinated to prevent an outbreak? You can estimate this using your simulation, or you can use Eq. (4.19) to calculate this exactly!

(d) Vaccines are often not 100% effective. For example, the annual flu vaccine is estimated to be around 60% effective. Repeat parts (a) and (b) assuming that the vaccine is 60% effective. How does this change your previous conclusions?

Exercise 4.22

(a) Instead of vaccinating only at the beginning of the season, we assume that the susceptible population is vaccinated at a rate of 1% of the population per day. Simulate the model for this scenario and plot $I(t)$. Compare your results to the case where no one is vaccinated.

(b) Near the end of the outbreak (around $t = 50$), what fraction of the population has been vaccinated? (Hint: We can create a new compartment $V(t)$ similar to what we did with the cumulative compartment $C(t)$!)

4.3.6 2009 H1N1 influenza pandemic

During late spring and early summer in 2009 public health officials all over the world were reporting a very new type of influenza. There were huge concerns since it did not match the previous influenza season vaccine, and it appeared that young adults, usually the most resilient to influenza, were much sicker than usual with high hospitalization and mortality rates. This put public health services on high alert. However, they did not want to needlessly sow panic or unnecessarily implement extremely disruptive measures (e.g., closing schools). They did however alert health care workers to report cases, so they could track this new disease, and many countries initially screened international airline passengers for fever to limit transmission across borders. Most countries collected data and shared information as it became clear that this was a true pandemic. One of the difficulties in collecting data on how many individuals have or have had influenza is the common nature of the disease itself. Even though this outbreak resulted in anomalously severe disease in young adults, many individuals had mild or even no clear symptoms at all. The symptoms (fever, congestion, and aches) occur with many diseases and do not last very long. Since influenza outbreaks occur annually, many people do not think it is worth going to the doctor. For all of these reasons, the numbers of cases in each setting are underreported to a degree that varies by community and country. Here we explore the pandemic as it developed in Australia with its very diverse population of 22.2 million people living mostly in large coastal cities in 2009.

Estimating parameters

In the following exercises, we will use data from the Australian Influenza Report (plotted in Fig. 4.11) to calibrate the SIR model.

Exercise 4.23

The data shown in Fig. 4.11 are weekly observed case incidence data. We can transform these data into cumulative number of infected cases, which can then be used to compare to our model output variable.

Input the data from Fig. 4.11:

```
t.data = seq(0,18)
new.cases.data = c(122, 722, 758, 832, 1064, 1505, 2520, 3610, 5188, 6069,
```

```
4479, 3891, 2741, 1860, 1052, 685, 477, 257, 192)
```

Many cases are not reported to surveillance systems, and therefore it is important to take this into account. We assume that only a proportion of infected individuals were reported. The reporting rate p_{report}, which incorporates both the rates of unreported symptomatic cases and asymptomatic cases, was estimated to be 0.0053 for Australia [47]. Use p_{report} to convert the reported cases defined as `new.cases.data` to the estimated actual number of cases and assign this to `adj.cases`. Then use these estimated actual numbers of weekly new cases to create a vector `C.cases`, which contains the cumulative number of cases. Use the command `cumsum` to help you do this!

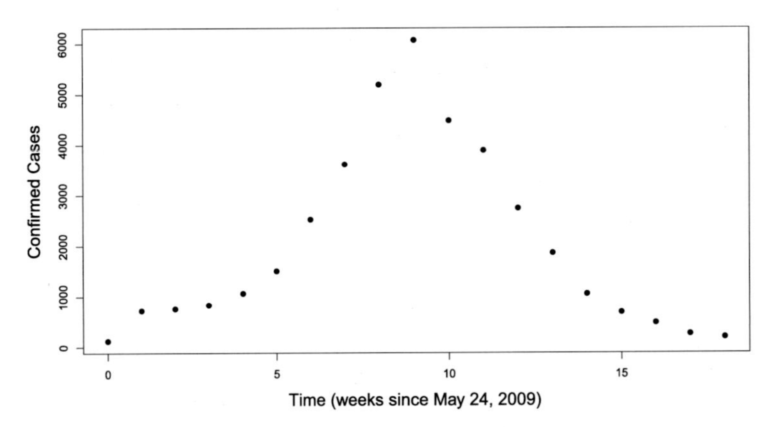

FIGURE 4.11

Confirmed H1N1pdm cases in Australia. Data extracted from Fig. 2 in the Australian Influenza Surveillance Report by the Australian Government Department of Health (http://www.health.gov.au/internet/main/publishing.nsf/Content/cda-ozflu-no2-10.htm).

Exercise 4.24

Now that we have estimated the cumulative cases by week, we can write a cost function that calculates the squared residuals between the model output and the data for a given set of parameter values. Then the optimization package `modFit` will call on this function to find the best-fit parameter values.

(a) Write a cost function called `SIR.cost` that has `parms` as input (see Listing 4.6 for an example). The `parms` vector will pass through values of b, k, and $I(0)$. Inside the cost function, call up your function `SIR`, which solves the differential equation (make sure you are using your modified function that includes $C'(t)$!). Then calculate the squared residuals between the model output $C(t)$ and `C.cases`. Make sure you are solving the differential equation system at the same time points as the data, that is, $t = 0, 1, 2, \ldots, 18$ weeks.

(b) Initial guesses for $I(0)$, b, and k need to be given as input to `modFit`. What can you use as an initial guess for $I(0)$? According to the Centers for Disease Control and Prevention (CDC), an influenza infection will typically last from three to seven days in most people. Based on this, what might you use for an initial guess for k? (Be careful! Previously, we used units of day^{-1}, but these data are in weeks.) For b, we know that the seasonal influenza usually has a basic reproduction number around 1.3; how can you use this number to estimate b?

(c) Use modFit to find estimates for b, k, and $I(0)$. See Listing 4.8 for an example.
(d) Plot the model output for $C(t)$ based on the parameter values you found in part (c) and overlay the data. How well does your model capture the dynamics throughout the epidemic?

Taking action

During the 2009 H1N1 pandemic, decision-making organizations such as the US Centers for Disease Control and Prevention (CDC) and the World Health Organization (WHO) invited outside modelers to assist in their response operations. Assume that it is the end of September, 2009, near the end of the outbreak in Australia (the right tail in Fig. 4.11). However, second waves are beginning to occur in places such as the United States and Canada. It is suspected that the virus is evolving and increasing its fitness in the human host, resulting in higher transmissibility. An organization in Australia has temporarily hired you to use a mathematical model to simulate various control strategies to assist them with their ongoing response operations.

Exercise 4.25

Based on your previous simulations of the outbreak that just ended, an estimated one-third of the population in Australia has been infected with H1N1. We will assume that if some individuals were recently infected with H1N1, they developed immune resistance to this new viral strain and will not be infected again. Therefore, the susceptible population is now approximately two-thirds of the total population.

In our simulations, we will assume that $k = 2.5$ per week and the transmissibility is projected to increase to $b \approx 2 \times 10^{-7}$. Also, we will assume that there are 2,270 individuals infected, that is, $I(0) = 2270$, which is what our current model predicted for the end of October. Based on these predictions, you need to investigate the extent of a possible second wave of the epidemic and develop strategies that might avoid the epidemic.

(a) Solve the SIR model using the given parameter values and initial conditions. Based on these values, will a second wave occur? If so, how many people will be infected during the second wave?
(b) A vaccine is now available for H1N1. Consider various vaccination strategies (vaccinating a fraction of the susceptible population at time zero or performing weekly vaccinations), keeping in mind that there will be a limit to how much vaccine will be available and a limit on facilities and people who can administer it. Include the fact that researchers have estimated the overall effectiveness of this vaccine to be 56%. Give a report to the organization based on your findings and include figures to support your findings.

Using this very simple version of the SIR model can quite readily reveal trends that are extremely useful to policy makers. However, there are many additional refinements to the model and other trends that might shape policy as well. Similarly, there are things that are nearly impossible to be sure about in the midst of an epidemic. For example, while the epidemic was on-going, the unusual virulence of this strain among young adults was detected. Did that also mean greater transmissibility or longer recovery rates for that age group? Since influenza is spread through close contact through virus particles in the air or left on surfaces due to sneezes and coughs, should we consider the differences in density or numbers of contacts that, say, school children or nursing home residents have versus the rest of the population? What percent of the population lives in rural areas? Would we expect the epidemic to grow differently in urban areas? What happens during school holidays? Finally, we must remember that influenza virus is famous for its ability to mutate in ways that avoid the immune system and that individual immunity to each strain of influenza virus wanes over time.

Choose just one of these complications and brainstorm about how you might change the fundamental SIR model or design your analysis to account for the actual substructure of the population and dynamics of the virus–human interactions.

4.4 Case study 2: Optimizing immunotherapy in prostate cancer

Goals: We will formulate a system of differential equations that describe the interactions among tumor vaccine cells, immune response, and prostate cancer cells. The effects of changing vaccine amount and injection schedules will be explored. Data from a prostate-specific antigen (PSA) test will be used to estimate the cancer cell proliferation rate and the tumor cell killing efficacy. The model will be applied to test vaccination schedules with the goal of stabilizing the number of tumor cells. In addition, sensitivity analysis will be used to identify model parameters that, if manipulated, are likely to make the greatest difference in cancer patient survival time, and we will simulate the effect of changing these parameters.

4.4.1 Background

Cancers are challenging to treat as they are our own cells that have lost control of their cell division rates as a consequence of accumulated mutations. Many treatments target rapidly dividing cells but are poorly tolerated as they also damage some healthy tissues. As cancers progress, mutations increase to the point that the immune system no longer recognizes them as "self". Once a tumor is antigenic, a set of immune system cells, collectively referred to as effector cells, will quietly kill only the cancer cells, often limiting cancer growth with no side effects. However, our natural immune response is self-limiting because activation against a target simultaneously initiates steps that will inhibit the effector cells. This down-regulation is sensible for fighting a transient infection but not helpful with persistently growing cancers. Moreover, many cancers create microenvironments that suppress the immune system. Clinicians hope to enhance targeted immune system responses by either interfering with the regulatory system or by boosting the presence of anticancer effector cells, especially the highly selective T lymphocytes. These are expensive approaches requiring complex protein-based drugs (e.g., an antibody to a down-regulatory cytokine) or the preparation of cancer-specific (often patient-specific) vaccines to stimulate the steps to activate targeted T lymphocytes. The challenge is to identify clinically appropriate strategies and then figure out working protocols (doses, timing) to improve patient outcomes, both quality of life and survival. Mathematical models have a key role in this research (e.g., [20]) due to the complexity of the system and the difficulties in obtaining key data, especially in human patients.

Prostate cancer is an intriguing target for immunotherapy because it is so common, and it is typical of other epithelial adenocarcinomas (e.g., skin cancer) suggesting that useful strategies might be broadly applicable. Prostate cancers are typically first treated by chemo- or radiation therapy, followed, if needed, by androgen ablation therapy since early tumors depend on androgens to grow. Prostate cancers can evolve to grow without androgens from nondetectable masses with no symptoms to overt disease; the median time to death for these patients is about 16 months [14]. The hope is that immunotherapy at this stage in the disease can limit growth and prolong high quality of life and survival.

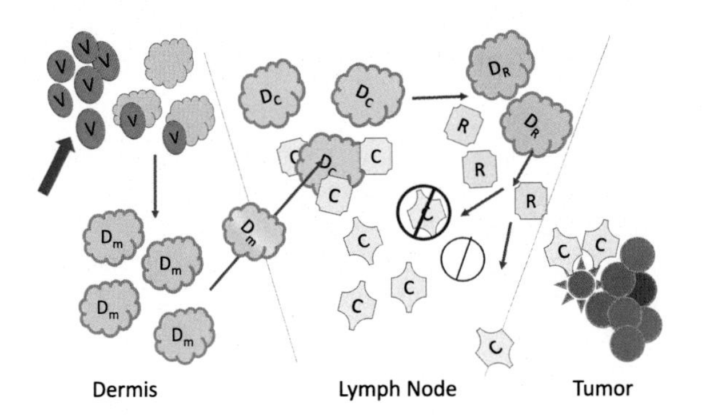

FIGURE 4.12

Cartoon of the cascading events initiated by vaccination with inactivated antigenic cells (V) that activate dendritic cells (D_m), which present the antigens and migrate to the lymph node as mature dendritic cells (D_c), which can activate cytotoxic T cells C targeting and killing tumor cells. They evolve to regulatory dendritic cells (D_R), which stimulate T_{reg} (R) to inactivate the cytotoxic T cells and thus limit the immune response.

Since the prostate tumor environment is poised to suppress the natural immune response, the strategy we will explore is enhancing immunological activity by vaccinating the patient with tumor antigens in the form of three types of inactivated allogenic prostate cancer cells (the vaccine) injected into the dermis. Vaccines initiate the sequence of events indicated in Fig. 4.12, in which cell–cell or cell–signal interactions initiate phenotypic changes, that is, changes in cell shape, receptors, and/or signal secretion that alter cell properties. Naive dendritic cells (DC) patrol the epithelial tissue where the vaccine is injected and will take up and process the antigens from the vaccine cells, thus initiating the steps toward their maturation. These recruited maturing DCs (D_m) migrate from the epithelium to the local lymph tissue. Once these DCs present the ingested antigens on cell surfaces and undergo several other phenotypic changes making them mature DCs (D_C), they can bind to and activate cytotoxic T lymphocytes (CTLs). The activated CTLs kill only cancer cells with matching antigens. The mature DCs change to a regulatory phenotype (D_R) in response to the evolving cytokine environment and now secrete additional cytokines that activate regulatory T cells (T_{reg}), which inhibit the CTLs. In the experiments we are examining the vaccination schedule and dosage, with the goal of optimizing the magnitude and duration of CTL activity and thus limiting tumor growth. Our mathematical model will help identify useful therapy logistics and identify other avenues for therapy such as lowering T_{reg} activation. Note in Fig. 4.12 that the biological system we are exploring involves three compartments and multiple cell types that affect the phenotypes or characteristics of neighboring cells through direct (binding) or indirect (cytokine signal) interactions.

Exercise 4.26

(a) Sketch the system by drawing compartments that represent the vaccine, DCs (all phenotypes), CTL, T_{reg}, and tumor cells using Fig. 4.12 as a guide. Include the flow rates and variables that will likely affect each rate.

(b) State in your own words what happens from vaccination to down-regulation of the system. What must be true for immunotherapy to work?

Unfortunately, immuno-suppression via the T_{reg} and the cancer microenvironment limit the duration of the vaccine effectiveness., and therefore the clinical approach is to revaccinate at monthly time intervals [35] to increase the chance of successful treatment. Success is defined as significantly slowing the cancer growth rate, but how do we know the growth rate of, as yet, undetectable cancer masses? Fortunately, the prostrate leaks a "signature" protein into the blood stream called prostate specific antigen (PSA), which is easily measured. The more prostate tissue (in this case, cancer cells), the more PSA in the blood. The rate of PSA concentration increase in the blood is proportional to the rate of prostate cancer growth, giving us a readily accessible proxy for cancer growth rate.

Since there are multiple cells to consider—the cancer itself, the vaccine cells, and five types of immune system cells (the three DCs, CTLs, and T_{reg})—we will explore the behavior of each interaction independently and then put them together. Spatial and temporal relationships are also critical and can be manipulated using the migration and conversion rate parameters. (Caution: this is a simplified system designed to highlight the key immune system parameters for prostate cancers: other immunotherapy models include other components.)

4.4.2 Model formulation

Modeling the immune response in a patient receiving vaccination therapy requires developing a system of mathematical equations that describe the dynamics of the vaccine, the immune response, the prostate cancer cells, and the interactions among these types of cells. Here we will build step by step a model of the cascade of events from vaccination to activated effector cell C-induced destruction of the prostate cancer cells P.

As discussed in the background, the vaccine injection stimulates maturation of naive DCs into antigen-presenting dermal DCs (D_m), and the vaccine is injected repeatedly. To start, we will model the dynamics of vaccine V-induced formation of D_m in the dermal compartment and explore how to simulate repetitive vaccinations.

Exercise 4.27

Dendritic cells respond to injected vaccine in the dermis

(a) Given an injection of vaccine cells at time zero, how would you expect the amounts of $V(t)$ and $D_m(t)$ to change over time? On the same plot, sketch $V(t)$ versus time and $D_m(t)$ versus time assuming, for now, that all the cells stay in the dermis. Note that the source of naive DCs is extremely high and constant, and therefore the disappearance of the vaccine is not limited by their availability. If this was not the case, then we would have to account for them in an equation for V.

(b) The differential equations for V and D_m can be given by

$$\frac{dV}{dt} = -k_i n_v V, \tag{4.20}$$

$$\frac{dD_m}{dt} = k_i V - k_m D_m, \tag{4.21}$$

where n_v is the number of vaccine cells required to induce maturation of one DC. Here we assume that $n_v = 1$ (and it is dimensionless), which means that each DC precursor takes up one vaccine cell to become an antigen-presenting DC. These antigen-presenting dermal DCs, D_m, then migrate to the lymph nodes at a rate proportional to D_m. Label the arrows in your sketch with appropriate rate constants. Give interpretations of the parameters k_i and k_m. (Note the similarity to the caffeine model in Unit 2!)

(c) Let $k_i = 1.4$ day^{-1}, $n_v = 1$, $k_m = 0.65$ day^{-1}, $V(0) = 2.4 \times 10^7$ cells, and $D_m(0)$=0. Solve the system of two differential equations (4.20) and (4.21) in R and plot $V(t)$ and $D_m(t)$ versus time on the same plot (see Listing 4.16). Do the results match your intuition? How might you use these results to develop the dose and frequency of a vaccine regimen? What other things would you need to consider?

Listing 4.16 (unit4ex16.R)

```
dermis = function(parms, times, V0, D.m0){
  with(as.list(c(parms)),{

    # Define differential equation
    dermis.model= function(t,y,parms){
      with(as.list(c(y)),{
        dV = -k.i*n.v*V
        dD.m = k.i*V - k.m*D.m
        list(c(dV,dD.m))
      })}

    # Solve system of differential equations
    y0=c(V=V0, D.m=D.m0)
    out=ode(y0,times,dermis.model,parms)
    as.data.frame(out)
  })
}

Vd=2.4e7 # Vaccine dose (number of cells)
n=1000 # Length of time vector
parms=c(k.i=1.4,n.v=1,k.m=0.65)
out = dermis(parms, times=seq(0,30,length.out=n), V0=Vd, D.m0=0)

PLOT THE NUMBER OF VACCINE CELLS AND DENDRITIC CELLS VERSUS TIME
```

(d) Simulate the model as you did in part (c) but now inject 2.4×10^7 vaccine cells again at $t = 30$. Note that you will need to solve Eqs. (4.20) and (4.21) twice! Solve the system as you did in part (c). Then use the values $D_m(30)$ and $V(30) + V_d$ (where V_d is the new injection of vaccine) as new initial conditions and simulate the model from 30 to 60 days (see Fig. 4.13).

(e) What if we want to inject the vaccine multiple times? (This problem is very similar to sipping coffee as we did in Exercise 2.40.) We want to make our code efficient and easy to manipulate later (e.g., vary dose, frequency of dose)! Since we will need to repeatedly solve the differential equation system, this is a perfect example of when a `for loop` is useful (refer to the R primer Section on Iteration 1.2.6). Assume that a patient receives an injection of 2.4×10^7 vaccine cells every 30 days for one year. Use the template in Listing 4.17 to simulate multiple doses using a `for loop`.

Listing 4.17 (unit4ex17.R)

```
t.inj = 30*(1:12) # 12 injections every 30 days

# Simulate model for first 30 days, t=0,...,30
out = dermis(parms, times=seq(0,t.inj[1],length.out=n), V0=Vd, D.m0=0)
out.total = out # Create out.total which will store output for all time

for (i in 1:(length(t.inj)-1)){
    INSERT INITIAL CONDITION FOR V0 AND D.m0
    out = tumor(parms, times=seq(INSERT INITIAL TIME, INSERT END TIME, length.out=n),
                V0, D.m0)
    out.total = rbind(out.total,out)  # rbind adds output as a new row to out.total
}
PLOT V(t) AND D.m(t), INCLUDE LEGEND
```

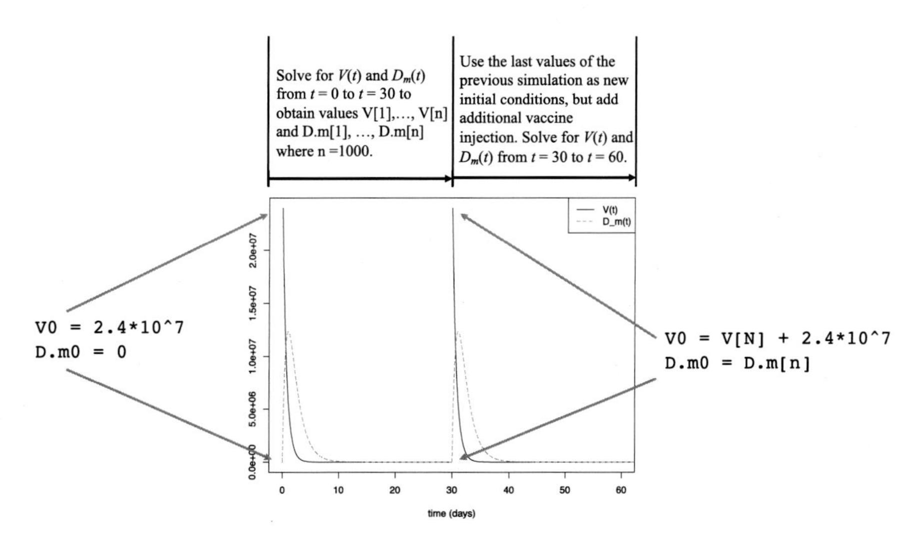

FIGURE 4.13

Implementing initial conditions when solving Eqs. (4.20) and (4.21) for vaccine injections at 0 and 30 days.

After the vaccine injection stimulates maturation of naive DCs into antigen-presenting DCs (D_m); some of these DCs migrate from the skin into the lymph node with a fraction a_l making it. These antigen-presenting DCs in the lymph node, denoted by D_C, recruit and activate tumor-specific CTLs (C) that kill tumor cells. However, the mature DCs in the lymph node also become "exhausted" (D_R) and secrete cytokines that recruit regulatory cells (R), which in turn inhibit CTLs. The populations of the two types of DCs, regulatory cells, and CTLs in the lymph node can be modeled by [35,50]

$$\frac{dD_C}{dt} = \alpha_l k_m D_m - k_{CR} D_C, \tag{4.22}$$

$$\frac{dD_R}{dt} = k_{CR} D_C - \mu_D D_R, \tag{4.23}$$

$$\frac{dC}{dt} = a_C D_C - \mu_C C - k_R C R, \tag{4.24}$$

$$\frac{dR}{dt} = a_R D_R - \mu_R R. \tag{4.25}$$

Exercise 4.28

Lymph node

Using the information in the background and the qualitative diagram in Fig. 4.12, explain what each expression represents in the right sides of Eqs. (4.22)–(4.25).

Finally, we consider how the number P of prostate cancer cells changes over time. The number of cancer cells changes due to their intrinsic growth/death rate r and increased death rate due to interaction with CTLs. Kronik et al. [35] propose the following model:

$$\frac{dP}{dt} = rP - a_P C P \frac{h_P}{h_P + P}. \tag{4.26}$$

where a_P is the maximal tumor killing efficiency by CTLs and h_p represents a damping coefficent.

Exercise 4.29

Tumor

(a) The growth rates of tumors from various patients were measured by Kronik et al. to obtain parameter estimates. The growth data did not saturate, and therefore they assumed the growth rate rP. What caution should you use when simulating the model in the long term and interpreting your results? (Hint: compare this formulation of tumor growth with those we used in the case on tumor growth in Section 3.5.)

(b) A mass action term, $a_P CP$, is one way to represent the interaction term between tumor cells and CTLs. Why not assume a simple one-to-one interaction? How does the factor $\frac{h_P}{h_P+P}$ represent what might really happen?

4.4.3 Model implementation

Before using patient data to estimate model parameters, we will use a given set of parameter values to verify that we can solve Eqs. (4.20)–(4.26) numerically in R and understand the model output.

Exercise 4.30

(a) Similar to the function `dermis` in Listing 4.16, create a function called `tumor` that includes the entire model given in Eqs. (4.20)–(4.26). Using the parameter values in Table 4.2, along with $r = 0.01$ and $a_P = 6 \times 10^{-7}$, solve Eqs. (4.20)–(4.26) for $t = 0, \ldots, 50$ with initial conditions $V(0) = 2.4 \times 10^7$, $D_m(0) = 0$, $D_C(0) = 0$, $D_R(0) = 0$, $C(0) = 0$, $R(0) = 0$, and $P(0) = 10^8$. Plot $P(t)$ and $D_m(t)$ on separate plots. Then show $D_C(t)$, $D_R(t)$, $R(t)$, and $C(t)$ on the same plot. Explain the dynamics of each cell type in terms of the immune response considering the shapes, magnitude, and persistence of each cell type. What happened to the tumor cell population following the vaccine injection?

(b) Now simulate the model using the following vaccination regimen: 2.4×10^7 cells are injected at times $t = 0$, $t = 14$ days, and $t = 28$ days and then every 28 days for a total of 14 injections. This is similar to what you did in Listing 4.17, but now you are solving the full system, and the time of injections are different. Plot $P(t)$ for $t = 0, \ldots, 360$ and check your result with the figure below.

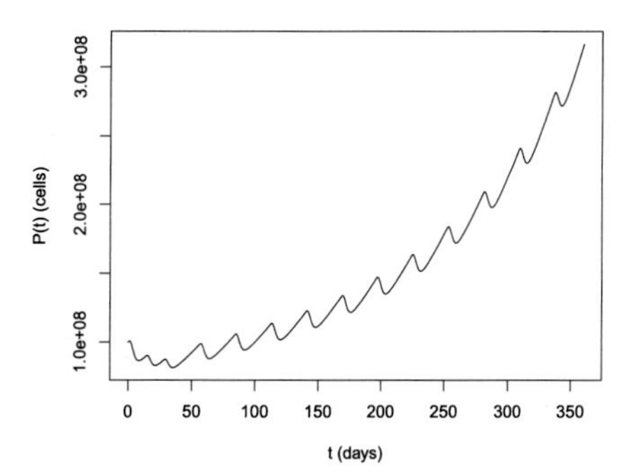

4.4.4 Parameter estimation

Most of the model parameters have been measured from various in vitro and in vivo studies and are listed in Table 4.2 [35,50]. However, it has been shown that the rate of tumor growth and the intensity of vaccine-induced immune response vary significantly among individuals [35], and therefore the tumor growth rate r and the CTL killing efficacy a_P must be estimated specifically for each patient.

Table 4.2 The parameters r and a_p are fit to patient data, and all other values are from [35] and [50].

Parameter	Definition	Value
k_i	Activation rate of mature DC by vaccine	1.4 day^{-1}
k_m	Migration rate of DCs from skin to lymph node	0.65 day^{-1}
α_l	Fraction of antigen-presenting DCs entering the lymph node	0.03
k_{CR}	Exhaustion rate of mature DCs	0.65 day^{-1}
μ_D	Death rate of exhausted DCs	0.092 day^{-1}
a_R	Activation rate of inhibitory cells by exhausted DCs	0.072 day^{-1}
μ_R	Death rate of inhibitory cells	0.72 day^{-1}
a_C	Activation rate of effector cells by mature DCs	0.4 day^{-1}
μ_C	Death rate of effector cells	0.17 day^{-1}
k_R	Inactivation rate of effector cells by inhibitory cells	1.4×10^{-5} $(\text{day} \cdot \text{cell})^{-1}$
r	Proliferation rate of tumor	Fit to patient data (day^{-1})
a_p	Maximal tumor cell killing efficacy	Fit to patient data $(\text{day} \cdot \text{cell})^{-1}$
h_p	Effector cell efficacy damping coefficient	10^8 cells

Here we use data from a patient enrolled in a study to test the effect of a prostate cancer vaccine. To be enrolled, the patients had to have progressed from a prostate cancer controlled by androgen ablation to reoccurring cancer indicated by an increased rate of PSA appearance in the blood. Recall that increasing PSA suggests growing prostate cancer even if it cannot be detected by mass or other symptoms. All patients meeting these criteria were given scheduled doses of the prostate cancer vaccine, and their PSA levels were routinely monitored. Data for a specific patient, given in Fig. 4.14, can be used to approximate r and a_P.

Exercise 4.31

If there is no vaccine treatment, the model assumes that the tumor cells grow exponentially, that is, $\frac{dP}{dt} = rP$, which implies $P(t) = P_0 e^{rt}$.

(a) Apply `nls` to the prevaccination data

```
t.prevac.data = c(0,  30, 70)
cells.prevac.data = c(1.46e8, 1.67e8, 2.05e8)
```

to estimate r. Report your value of r, plot the data, and model output.

(b) Based on your estimate for r, how long would it take for the number of tumor cells to double if no treatment is used?

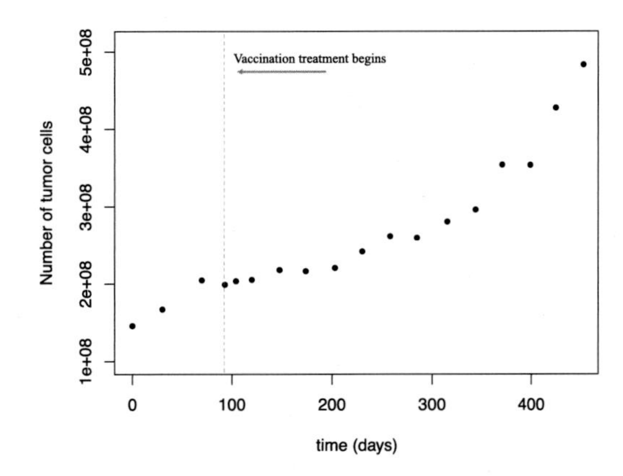

FIGURE 4.14

PSA measurements were extracted from Fig. 2 (Patient 14) in [35]. Then, the linear equation Volume (cm^3) $= 3.476 + 0.302 \times$ PSA (ng/l) [32] was used to convert PSA measurements to tumor volume in cubic centimeters. Assuming that one tumor cell is approximately 2.7×10^{-8} cm^3, the volume was then converted to the number of cells.

Exercise 4.32

To estimate a_P, the maximal tumor killing efficiency, least squares can be used to find the value of a_P that minimizes the squared residuals between the model output for P (simulated from Eqs. (4.20)–(4.26)) and the data following the initiation of treatment.

In this exercise we will first create a cost function that computes the squared residuals between the model output and data for a given value of a_P. Then we will use the built-in optimization function `modFit` to find the value of a_P that minimizes the cost function.

(a) The function `modFit` calls up a function (a "cost" function that we need to create!) that calculates and returns a vector of squared residuals. We must solve the differential equations (4.20)–(4.26) using the vaccine injection schedule as described previously—injections at days 0, 14, 28 and every 28 days thereafter for a total of 14 injections. The cost function is given in Listing 4.18. Look closely at each line and make sure you understand the code. Note that it uses the `tumor` function you created in Exercise 4.30. Once the equations are solved, we need to select the model output for P at the time points at which the data are obtained. Look up the commands `which` and `%in%` and explain what these commands do. Also, explain what `P.mod` is and how it is obtained.

Listing 4.18 (unit4ex18.R)

```
# Data, t=0 corresponds to the time of the first injection of vaccine cells
t.vac.data = c(0, 11, 27, 55, 81, 110, 137, 165, 192, 222, 250, 277, 306, 332, 360)
cells.vac.data = c(2.00e8, 2.03e8, 2.05e8, 2.18e8, 2.16e8, 2.20e8, 2.42e8, 2.62e8,
                   2.60e8, 2.81e8, 2.96e8, 3.54e8, 3.54e8, 4.27e8, 4.83e8)

# Cost function - calculates squared residuals between data and model output P(t)
tumor.cost = function(parms){
```

```
  with(as.list(c(parms)),{
    parms=c(k.i=1.4, k.m=0.65, alpha.1=0.03, k.CR=0.65,
            mu.D=0.092, a.R=0.072, mu.R=0.72, a.C=0.4, mu.C=0.17,
            k.R=1.4e-5, r=0.005, a.P=a.P, h.P=10^8)
    Vd=2.4e7    # number of vaccine cells injected

    # Time of injections. Last entry is not an injection, but gives final time.
    t.inj = c(14, 28, 56, 84, 112, 140, 168, 196, 224, 252, 280, 308, 336, 364)

    # Solve system for t=0,...,14. Calls up function 'tumor' where differential
    # equation is defined and solved
    V0=Vd; D.m0=0; D.C0=0; D.R0=0; C0=0; R0=0; P0=cells.vac.data[1]
    out = tumor(parms, times=seq(0,t.inj[1],by=0.1), V0, D.m0, D.C0, D.R0, C0, R0, P0)
    out.total=out

    for (i in 1:(length(t.inj)-1)){
      n = length(out$time)
      V0=out$V[n]+Vd; D.m0=out$D.m[n]; D.C0=out$D.C[n]; D.R0=out$D.R[n]
      C0=out$C[n]; R0=out$R[n]; P0=out$P[n]
      out = tumor(parms, times=seq(t.inj[i],t.inj[i+1],by=0.1), V0, D.m0, D.C0,
                  D.R0, C0, R0, P0)
      out.total = rbind(out.total,out)  # Add new output to out.total
    }

    P.mod = out.total$P[which(out.total$time %in% t.vac.data)]

    # Calculate squared residuals
    return((P.mod-cells.vac.data)^2)
  })
}
```

(b) Use `modFit` to find a_P. Recall that you will need an initial guess for the value of a_P. Think about a reasonable order of magnitude for this parameter. You can also use `manipulate` to try different values of a_P and compare to the data.

(c) Using the value of a_P obtained in part (b), plot the data, and model output for $P(t)$.

4.4.5 Vaccination protocols and model predictions

Vaccination protocols can be considered using the same framework as drug administration to stabilize the effect by varying the frequency and amount of drug per dose. Here the goal is to optimize and stabilize the ability of the CTLs to kill the cancer cells without harming the patient, which is similar to determing an optimal ibuprofen dose and schedule to minimize pain.

Recall that the purpose of this treatment is to limit the growth rate or even reverse the prostate cancer cell population. Not every patient will respond the same way. For example, larger cancers are harder to treat. Differences in immune system response at any one of the steps will lead to variations in patient outcomes. The first two things to try are to match the vaccine dose and scheduling to the needs of the patient. First, we need to understand how varying these two conditions will change the size of the prostate tumor cell population P.

Exercise 4.33

(a) Using the parameter values from Table 4.2, along with $r = 0.005$, $a_P = 3.8 \times 10^{-7}$, and $P(0) = 2 \times 10^8$, try varying the amount of vaccine given per treatment. Think about what might be reasonable to inject; for example, increasing the amount 10-fold might not be tolerated by the patient. Plot your results in terms of the number of prostate tumor cells P over time for 12 injections at the new dose. Did anything change significantly? If not, increase (or decrease) the dose. Does increasing dose size seem like a good strategy?

(b) Explore the effect of changing the vaccination schedule. The standard maintenance schedule is once every four weeks. Try one or two different schedules for a year and plot the number of tumor cells over time. What is the effect? Explore the process as you did previously after the first and last doses given. Which steps in the process are responsible for the effects you see?

4.4.6 Sensitivity analysis

One problem with just adding more vaccine more frequently or more vaccine per dose is that the down regulatory arm of the immune response is being activated at the same time as the desired activation of CTLs. Refer to your results in Exercise 4.33. Does "more" always work? Fortunately, in a system with this many steps, it is likely that there are additional therapeutic opportunities by disrupting or enhancing steps in between the initial activation of the DCs by the vaccine and the CTL effects on P.

How can you figure out which steps are most sensitive to an increase in vaccine dose? Is there a step in this system that might limit the effectiveness of increasing the dose or amount of vaccine added? Using the previous exercise, you could plot the trajectories of each cell type at the first vs. last vaccination. Based on this information, which steps in the process are most affected by the increased vaccine dose or frequency? Which steps limit the effectiveness of increasing the dose or frequency? These observations could be used to generate hypotheses about what to do next therapeutically. However, this is a messy graph! What might a mathematician do? Sensitivity analysis will reveal the relative contributions of each parameter and thus identify logical places to perturb the system.

Exercise 4.34

Use the parameter values from Table 4.2, along with $r = 0.005$, $a_P = 3.8 \times 10^{-7}$, and $P(0) = 2 \times 10^8$, and assume that 2.4×10^7 vaccine cells are injected once at $t = 0$. Simulate the model for 30 days to compute the number of prostate cancer cells on day 30. If we want to understand how changes in each parameter affect the number of cancer cells on day 30, then we can perform a sensitivity analysis!

(a) Create a bar plot (similar to Fig. 4.3) that shows the percent change in the number of prostate cancer cells on day 30 when each parameter is increased by 1%.

(b) A large range of values for the tumor growth rate r and cell killing efficacy a_P have been found among patients [35]. Knowing this and given the sensitivity results, discuss why we might use caution in using this particular model as a tool to implement a vaccination regime for some general population of patients.

(c) Aside from r and a_P, what parameters have the greatest effect on the size of the tumor?

4.4.7 Simulating other treatment strategies

Each of the parameters in the model reflect the actual kinetics of the system. For example, the extent to which the mature DCs (D_c) are able to activate the effector cells depends on the cytokine or chemical environment in the lymph node and tumor. This can be manipulated most commonly with IL-2. Similarly, the entire down regulation system is mediated through multiple cytokines (e.g., TGF-β). The strength of the CDL interaction with the tumor cells depends on the vaccine antigenicity or its ability to induce an immune response, which can also be altered. What should a clinical researcher do next?

Exercise 4.35

(a) Based on your sensitivity analysis, choose one strategy to test with our model using the standard treatment protocol by adjusting the value of the parameter affected by your treatment option. Plot the output after one and after 10 treatments while varying the parameter of your choice.

(b) Prepare a short report on the potential efficacy (ability to stabilize or decrease the cancer) associated with manipulating your parameter. Include suggestions for the extent of change needed to have a threshold (e.g., 10% change in the rate of tumor growth).

4.5 Case study 3: Quorum sensing

Goals: We will explore how bacteria or other types of individual cells can assess their own population density and respond by switching phenotype when a quorum is present by using a system of differential equations to model sequential events and positive feedback. We will consider the bacterial population growth and the intracellular events (signal synthesis, receptor binding, gene regulation response) to model a temporal response and compare the behaviors of the variables. Steady-state data will allow us to estimate some of the model parameters, whereas other parameter values are obtained from previous studies. Global sensitivity analysis will reveal critical parameters that affect the switch-like behavior in the system. We will also explore how a bacterial population might adjust the timing of its response.

4.5.1 Introduction

The discovery that individual cells like bacteria can function as a group once their population hits a given density was made in one of the most enchanting biological phenomena, marine luminescence by bacterial symbionts living with larger marine species. The tiny Hawaiian bob-tail squid (Fig. 4.15) harbors *Aliivibrio fischeri* a Gram-negative marine bacterial species that inhabits a special pouch, the squid's light organ. This helps the squid by providing a "cloak" of blue light matching moon light penetrating through the water so that the nocturnal squid is hidden from predators swimming below. However, at dawn the light turns off, and during the day, these tiny squid burrow in the sandy ocean bottom to sleep and hide. How is the light turned on and off?

Bob-tail squids control the *Aliivibrio* population density in their light organs by eliminating at least 95% of their bacteria just before morning; the remaining bacterial population grows at least 20-fold during the day. Could the ability to emit light be a density-dependent phenomenon? Early experiments demonstrated that it is, with light emission appearing to turn on like a switch, that is, light emission occurs at a threshold bacterial density, a quorum, and rapidly reaches its maximum response.

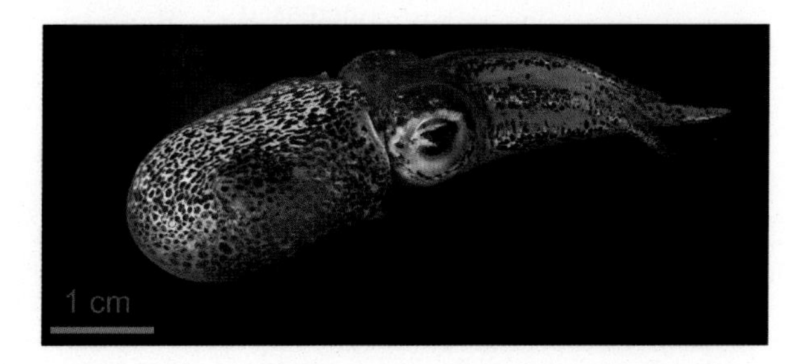

FIGURE 4.15

Hawaiian bobtail squid. Photo by Margaret McFall-Ngai – Divining the Essence of Symbiosis: Insights from the Squid-Vibrio Model, CC BY 4.0, https://commons.wikimedia.org/w/index.php?curid=31023377.

Quorum-based phenomena have been found throughout the bacterial kingdom and include initiation of biofilm formation (when bacteria organize into a colony on a surface), upregulation of virulence factors, and secretion of extracellular enzymes needed for nutrition, behaviors that are only feasible with a dense bacterial population. To be efficient, the genes for these expensive activities are switched off unless there is a quorum. Constitutively sending a small molecular signal (the autoinducer) at low levels from each individual is an inexpensive way of indicating bacterial density. Bacteria monitor the concentration of autoinducer via internal receptors, which "decide" whether or not to invest in the expensive proteins needed to engage in group behavior. How does this work? The absolute concentration of the signal will increase if there are more bacteria secreting it in a small volume (draw this for yourself). These signals are transcription regulators that ultimately initiate new gene expression leading to the proteins needed for luminescence (or biofilms).

Quorum sensing requires a number of sequential steps: a secreted signal, a receptor, and a means of upregulating the response genes. Most members of the bacterial population send, receive, and respond to the signals. Gram-negative bacteria like *Aliivibrio fischeri* use the *lux* genes for quorum sensing with the downstream response being luminescence or other types of responses such as biofilm formation, depending on the species. These components are well known (Fig. 4.16). The autoinducer signal is a small membrane permeable molecule, an acyl-homoserine lactone (AHL) (e.g., 3-oxo-hexanoyl-homoserine lactone) synthesized by the LuxI enzyme, the gene

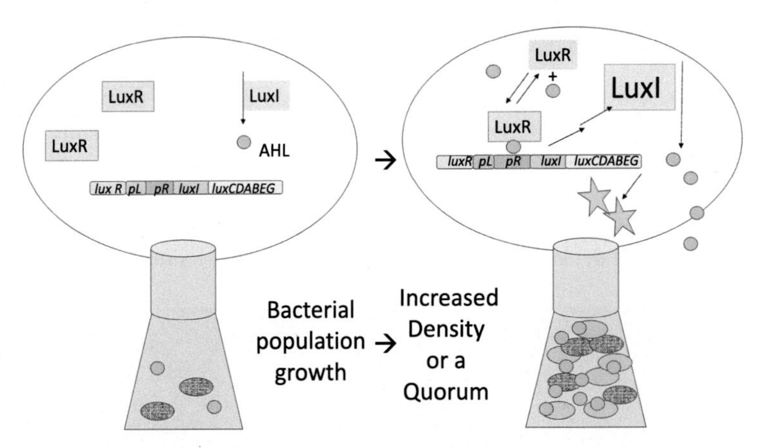

FIGURE 4.16

Diagram of the lux system for quorum sensing and its dependence on bacterial population density. Individual bacteria are represented by the large ovals. Green circles represent the autoinducer, AHL synthesized by the enzyme LuxI. The LuxR-AHL complex activates the promoter, upregulating transcription of the *luxI* gene and the genes for producing luminescence, *luxCDEAB*. The yellow stars indicate luminescence. The two flasks represent the cell culture with few bacteria and low concentrations of AHL initially. As the bacteria divide, density increases, AHL concentration increases and the number of luminesing bacteria (yellow) increases.

product of *luxI*. AHL diffuses from the cell into the medium; when the bacteria are dilute (e.g., free swimming in the ocean), AHL concentration is very low. Bacterial cells also make an AHL receptor protein known as LuxR coded by the *luxR* gene. AHL binds to LuxR forming $(LuxR\text{-}AHL)_n$, which binds to the promoter region (lux box) controlling the transcription of the luminescence genes (*luxCDABEG*) that code for the proteins required to generate light. The same promoter increases transcription of the *luxI* gene resulting in more of the enzyme LuxI and thus increased rate of AHL synthesis, that is, a positive feedback loop and possible switch mechanism. Is this how the control system actually works?

To model the entire system, we can start with what we know about logistic growth in bacteria. Then we need to model the regulatory system. What do we know? There is a constitutive low concentration of LuxI in cells. The promoter for *luxR* is controlled by a variety of agents generating a steady-state concentration of the receptor LuxR per cell. Central to the quorum response is the promoter that is activated by the LuxR-AHL complex controlling both the synthesis of *luxI* and thus the amount of the enzyme LuxI and the suite of genes needed for luminescence. As in all biological systems, each component has a unique degradation rate contributing to the overall system dynamics.

Exercise 4.36

(a) Using Fig. 4.16 as a guide, draw for yourself the components of the system indicating the relationships among them. List factors that control the concentrations of the autoinducer signal (AHL), the enzyme that forms it (LuxI), the receptor concentration (LuxR), and the complex that binds to the promotor.

(b) As we begin developing our model, we should impose some assumptions. Discuss how the following might be useful:

- AHL rapidly equilibrates between intracellular and extracellular spaces.
- The AHL LuxR interactions can be assumed to be approximately at steady state (recall substrate enzyme interactions).
- Only one complex can bind to the *lux* box or promoter for *luxI* and the luminescence genes.
- As cells grow and divide, their internal protein concentrations (e.g., LuxR) remain constant.

To explore the system completely, we want to consider the bacterial population growth rate, the volume of the vessel in which they are growing, and the responses as a function of AHL concentration. If we assume that the entire population responds to the regulatory system as a single bacterium would, then it is reasonable to focus our model on the regulatory mechanism itself. For example, what is the system dependence on AHL concentration?

4.5.2 Model formulation

To help us understand how bacterial population growth could result in bioluminescence and what steps affect when bacteria are "on" or "off", we introduce the following simple model describing the dynamics of the bacterial population and the

AHL, LuxR, and AHL-LuxR concentrations:

$$\frac{dN}{dt} = rN\left(1 - \frac{N}{K}\right), \tag{4.27}$$

$$\frac{dA}{dt} = \left(\alpha_A + \beta\frac{C^n}{C_{th}^n + C^n}\right)N - k_1 A R_0 + k_2 C - \gamma_A A, \tag{4.28}$$

$$\frac{dC}{dt} = k_1 A R_0 - k_2 C, \tag{4.29}$$

where $A(t)$ is the concentration of AHL in the system at time t, R_0 is the intracellular concentration of luxR (assumed to be constant), and $C(t)$ is the intracellular concentration of the AHL-LuxR complex at time t. We could distinguish between intracellular and extracellular concentrations of AHL. However, AHL diffusion into and out of the cell is assumed to be fast compared to the time scale of the other steps in the *lux* system, making it reasonable to assume that the intracellular and extracellular concentrations are in equilibrium.

The bacteria are assumed to follow simple logistic growth with N representing bacteria/mL, r the growth rate constant, and K the carrying capacity of the system, as we have seen before. AHL concentration will change due to synthesis, degradation, binding with LuxR, and dissociation of the AHL-LuxR complex. There are two terms describing the synthesis of AHL. Even if the lux box is unoccupied, there is a basal rate of expression represented by $\alpha_A N$. AHL production is also regulated by the AHL-LuxR complex, which when bound to the lux box with a stoichiometry of n, positively influences AHL production by increasing production of its synthase, LuxI. The function

$$f(C) = \frac{C^n}{C_{th}^n + C^n} \tag{4.30}$$

represents the proportion of time during which the lux box is occupied by the AHL-LuxR complex C, and C_{th} represents the threshold value for AHL-LuxR complex in the AHL production process. Multiplying $f(C)$ by the maximum production rate β, and the number of bacteria N gives the rate of AHL production regulated by the complex.

Exercise 4.37

(a) Eq. (4.30) has appeared in previous sections (see Exercise 3.5 in Unit 3, Holling Type III). Sketch the general shape of $f(C)$ and label C_{th} on your graph. Discuss how increasing n changes the shape of the curve.

(b) Why is a Hill coefficient n included in the complex-dependent AHL production term?

(c) Explain why both production rates for A contain N.

(d) Recall that AHL and LuxR bind to form the AHL-LuxR complex, and the complex can also dissociate. Which terms represent these processes?

(e) AHL naturally degrades, and this abiotic degradation rate can vary due to pH and temperature. Which term corresponds to the abiotic degradation of AHL?

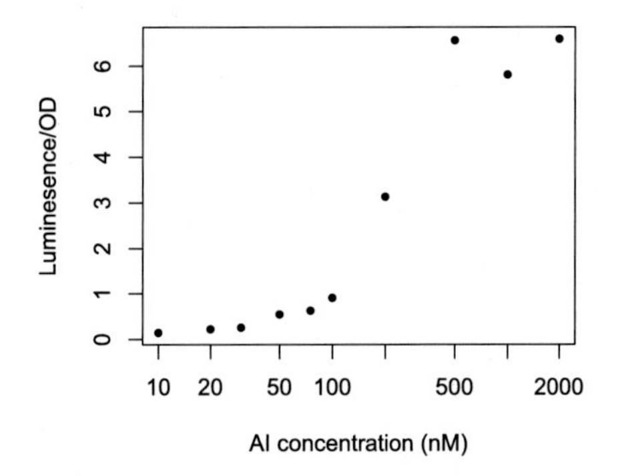

FIGURE 4.17

Aliivibrio fischeri luminescence/OD measured after 130 minutes incubation with exogenously added AHL from 10–2000 nM (note the log scale). Data extracted from Fig. 1 in [52].

4.5.3 Parameter estimation

Many of the model parameter values can be taken from various studies on *Aliivibrio fischeri* and other bacteria including *Pseudomonas aeruginosa* and *Pseudomonas syringae* (see Table 4.3). However, the parameters C_{th} and n are assumed to be unknown. Although time series data on AHL and AHL-LuxR concentrations are difficult to obtain, studies have measured luminescence at varying defined AHL concentrations (see Fig. 4.17 for an example). In this experiment, aliquots of exponentially growing *A. fischerii* were added to growth media containing the indicated amounts of N-3-oxyhexanoyl-L-homoserine lactone (the AHL for this system), thus setting the AHL concentration. The cells were incubated with the exogenous AHL at room temperature for 130 min, long enough for the luminescence response to reach a steady state. (Think about why it takes this long!) Luminescence is the light emitted by the bacteria, and the optical density (OD) reflects the number of bacteria per mL (recall Bacterial Growth lab in Unit 2) so that the y-axis in this graph is reflective of the per capita luminescence.

How can we use this data set to estimate C_{th} and n? Recall that $f(C) = \frac{C^n}{C_{th}^n + C^n}$ describes the proportion of time during which the lux box is occupied by the LuxR-AHL complex C. Therefore $f(C)$ represents the fraction of bacteria that upregulate the genes required for bioluminescence. Then $f(C) \cdot N$ is the number of bacteria that are luminescent, and the number of photons emitted per second by the bioluminescent colony is simply proportional to the number of bioluminescent bacteria $f(C) \cdot N$. This implies that the data for luminescence per cell in Fig. 4.17 can be fitted to the curve $g(C) = b\frac{C^n}{C_{th}^n + C^n}$, where b is a constant. Note that the independent variable in Fig. 4.17 is AHL concentration, but our function $g(C)$ is a function of C. However, A and C are related. Assuming that A and C are held constant (as they are in this steady-state experiment), their derivatives with respect to time are zero. Setting $\frac{dC}{dt} = 0$ implies $C = \frac{k_1 A_e R_0}{k_2}$, where A_e denotes the equilibrium concentration.

Replacing this expression for C in the function $g(C) = b\frac{C^n}{C_{th}^n + C^n}$, we obtain

$$g(A_e) = b\frac{\left(\frac{k_1 A_e R_0}{k_2}\right)^n}{C_{th}^n + \left(\frac{k_1 A_e R_0}{k_2}\right)^n}, \tag{4.31}$$

where g is now a function of the steady-state AHL concentration A_e. Multiplying the numerator and denominator by $\left(\frac{k_2}{k_1 R_0}\right)^n$, this becomes

$$g(A_e) = b\frac{A_e^n}{\left(\frac{k_2 C_{th}}{k_1 R_0}\right)^n + A_e^n}, \tag{4.32}$$

which can be used to find C_{th} and n, as described in Exercise 4.38.

Exercise 4.38

Eq. (4.32) can be written as

$$g(A_e) = b\frac{A_e^n}{K^n + A_e^n},$$

where $K = \left(\frac{k_2 C_{th}}{k_1 R_0}\right)$. We can fit this equation to the data in Fig. 4.17 using tools from Unit 3.

(a) Looking at Fig. 4.17, what are your estimates for K and b? Explain why you chose these estimates.

(b) Input the data in R:

```
AHL = c(10, 20, 30,  50, 75, 100, 200, 500, 1000, 2000)
Lum = c(0.143, 0.222, 0.257, 0.545, 0.626, 0.906, 3.130, 6.560, 5.810, 6.590)
```

Use your estimates from part (a) and choose a value for n as input to `nls` to fit the function $g(C)$ to the data. Graph the data and $g(C)$ on the same plot. Use a log scale for the x-axis, which can be done by including the option `log="x"` in the plot command.

(c) Using your results from nonlinear regression and the values of k_2, k_1, and R_0 in Table 4.3, report your estimates of C_{th} and n. (You should obtain $C_{th} \approx 2.5 \times 10^{-8}$ M.)

4.5.4 Model simulations

Now that the model is formulated and parameter values are estimated, we can simulate the model to look at the dynamics of each variable.

Exercise 4.39

(a) Solve Eqs. (4.27)–(4.29) using the parameter values in Table 4.3 and the values for C_{th} and n obtained in Exercise 4.38. Plot $N(t)$, and on a separate graph, plot both $A(t)$ and $C(t)$.

Table 4.3 Estimates taken from [18,52,71,56,11].

Parameter	Definition	Value
$N(0)$	Initial bacterial density	5×10^{10} cells/L
$A(0)$	Initial AHL concentration	10^{-9} mol/L
$C(0)$	Initial complex concentration	0 mol/L
R_0	LuxR concentration (assumed constant over time)	3.5×10^{-9} mol/L
r	Bacteria reproduction rate	0.5 h^{-1}
K	Bacteria carrying capacity of environment	10^{13} cells/L
γ_A	AHL degradation rate	0.8 h^{-1}
k_1	Binding rate constant between AHL and LuxR	2.4×10^8 L/(mol h)
k_2	Dissociation rate constant AHL-LuxR complex	6.7 1/h
α_A	AHL basal production rate	1×10^{-20} mol/(cell h)
β	AHL-LuxR-dependent production rate	2×10^{-19} mol/(cell h)
C_{th}	Threshold value for AHL-LuxR (in AHL production)	Fitted
n	Polymerization degree for AHL-LuxR (in AHL production)	Fitted

(b) Studies have shown that AHL approaches a steady-state concentration in the range 100–1000 nM. What does your model predict for the steady-state concentration of AHL?

(c) Note that the magnitude of values for A and C are different from the values of N by many orders of magnitude given the units we are using! To compare the time course of the bacteria population to that of the AHL concentration and AHL-LuxR concentration on the same graph, we can normalize the variables by dividing each variable by its maximum variable value. For example, if the output of A is given by `out$A`, then this can be normalized by creating a new variable given by `out$A/(max(out$A))`. This is a very handy strategy! Plot the normalized variables for bacteria population, AHL concentration, and AHL-LuxR concentration on the same plot (see Fig. 4.18).

(d) Studies suggest that AHL reaches its maximum value approximately 2 hours after the bacteria have reached their carrying capacity. Based on the graph you just plotted, what is the concentration of AHL during the mid-exponential phase of bacterial growth? When do they reach the carrying capacity? How long after reaching carrying capacity does AHL concentration reach a steady state? Do these temporal relationships between the behavior of N and A make sense to you? What about the shapes of the curves?

(e) Recall that the fraction of bacteria upregulated is given by $f(t) = \frac{C(t)^n}{C_{th}^n + C(t)^n}$. Plot $f(t)$ and interpret the result in terms of the bacteria expressing bioluminescence.

Another way to visualize the relationships among the variable solutions is plotting them in a phase plane. For example, given $N(t)$ and $A(t)$, we can start from the pair $(N(0), A(0))$ and plot the values in the NA-plane (N is on the horizontal axis, and A is on the vertical axis) at a given time. As t increases, $(N(t), A(t))$ sweeps out a curve in the NA-plane, which corresponds to the trajectory of the solution (Fig. 4.19).

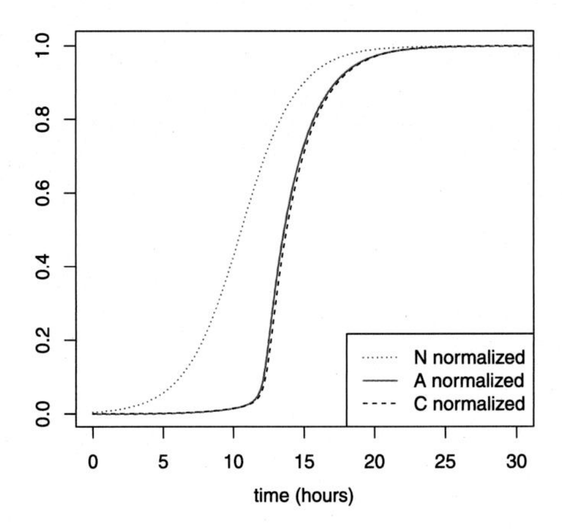

FIGURE 4.18

Bacteria population and concentrations of AHL and AHL-LuxR complex (normalized by their respective maximum values).

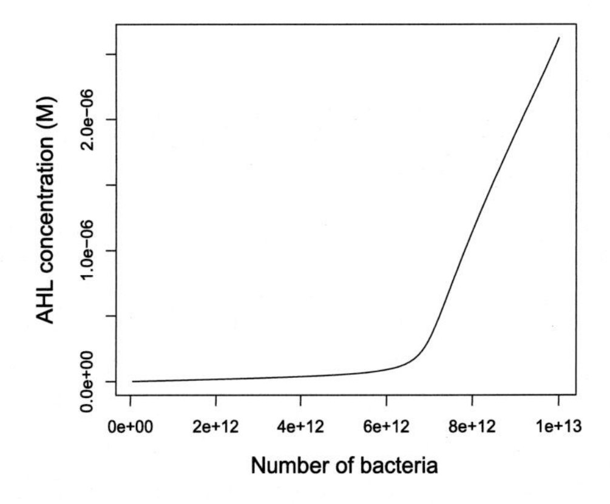

FIGURE 4.19

AHL concentration versus the respective number of bacteria N in cells/L shows that AHL begins to rise at a threshold bacterial density and rises steeply even as the bacterial numbers approach a steady state value.

Exercise 4.40

(a) Use your solution output from the previous exercise to plot $A(t)$ versus $N(t)$. Check your result with Fig. 4.19.

(b) What point on the curve corresponds to $t = 0$? What point corresponds to the final time point? If you were to draw arrows on the curve representing traveling forward in time, which way would these arrows point?

(c) Interpret your plot in terms of quorum sensing. How does the relationship between the bacterial population density and the AHL concentration help us think about the factors needed for the bacteria to "turn on."

4.5.5 Sensitivity analysis

To explore our simple model of the *lux* system, we depend on many parameters that were estimated from bacteria other than *Aliivibrio fischeri*. Moreover, a range of values has been reported for many of the parameters requiring us to make rough approximations. How do we know if the model we are exploring is remotely realistic in a qualitative or quantitative sense? This is one purpose of sensitivity analysis. Additionally, sensitivity analysis can reveal which parameters have the greatest effect on the model output. This latter information helps identify biological options (perhaps not yet identified) for modulating the quorum sensing pathway to adjust to current circumstances or that could evolve to serve different purposes. We will begin by perturbing each parameter by 1% and see if it alters the steady-state values of AHL and luminescence. Then we will explore more extensive changes in some of the parameters to get a better feel for the resilience of this system.

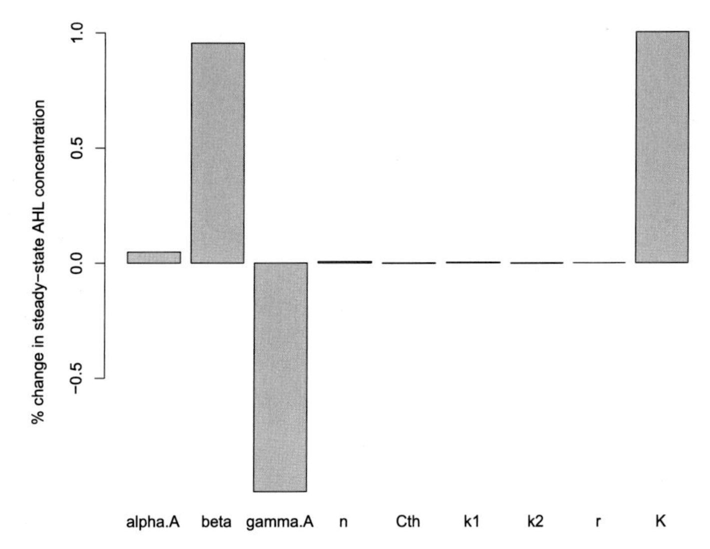

FIGURE 4.20

Effect of increasing each parameter by 1% on the steady-state AHL concentration.

Exercise 4.41

(a) Increase each parameter by 1% one at a time and calculate the percent increase in the steady-state AHL concentration. Display your results in a bar chart (refer to a similar example in Listing 4.12). Check your results with Fig. 4.20.

(b) Discuss your results in terms of the biological interpretation of each parameter. Specifically, the AHL steady-state concentration is most sensitive to which parameters? Examine the parameters that seem to have very small effects on the output in terms of the steps in the model they affect. Does low sensitivity to these make sense?

(c) Discuss how these results might help with future experimental design.

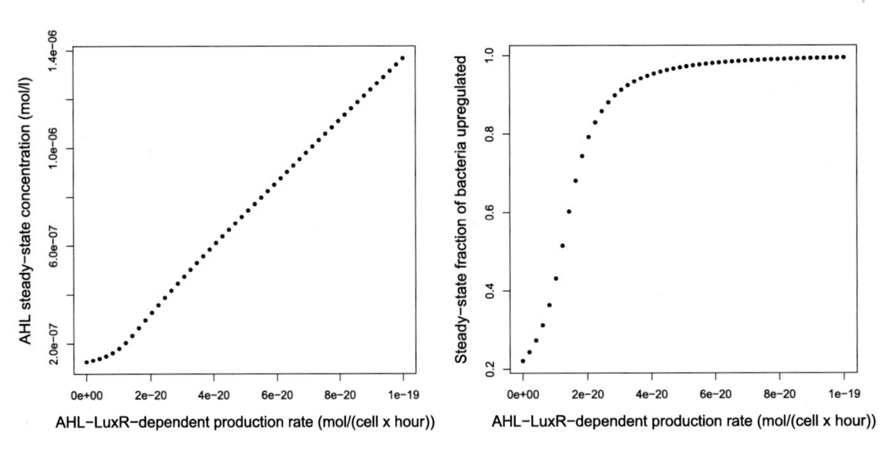

FIGURE 4.21

Steady-state values as a function of AHL production rate β.

Here we examine three of the factors that alter the steady-state concentration of AHL more closely to understand the dynamics of this system a bit better. One parameter that seems important is β, the rate constant for AHL formation if the system is fully upregulated to express the maximum amount of LuxI, the synthase for AHL. Here we have it set as 20-fold greater than the constitutive rate constant α_A. Do we trust our value for β? You may have noticed that many of the parameter values are tiny so that getting estimates requires clever experiments. One way of knowing how much new protein is being made when the *lux* is activated is measuring upregulation of a protein that is easy to measure, for example, one with a fluorescent tag. The maximum change appears to be 20-fold, thus setting an upper limit for our value of β. We might explore β more thoroughly by comparing the steady-state AHL concentrations achieved over a large range of values for this parameter (e.g., 0 to 2.0×10^{-19} mol/cell-hour) and compare that with the fraction of bacteria that are luminescent over the same range (Fig. 4.21).

Exercise 4.42

(a) Using Listing 4.11 as a template, plot the AHL steady-state concentration and steady-state fraction of upregulated bacteria as shown in Fig. 4.21. To do this, solve the differential equation for `times=seq(0,100,by=.1)` (AHL reaches a steady state within 100 minutes) for β values defined by `beta.vals = seq(0,1e-19,length.out=50)`.

(b) Carefully examine the two plots in Fig. 4.21 and note the value for β we used initially to solve the model. What is the minimum value of β required for significant numbers of bacteria to ultimately become luminescent? Why is the effect of changing the value of β on the two steady states so different? (Hint: look at the values on the y-axis of each plot.)

(c) What assumption(s) are we making that might affect these results?

Another key parameter value seems to be the AHL degradation rate constant γ_A. We chose the value 0.8 hr^{-1} from a wide range of observed natural degradation rates

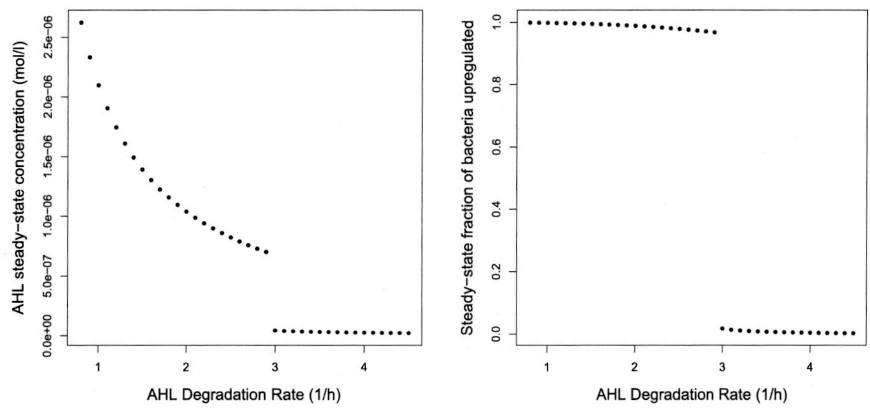

FIGURE 4.22

Steady-state values or AHL concentration and luminescence as a function of AHL degradation rate.

in which the AHL concentration in bacteria-free media is measured over time. The rate is pH and temperature dependent. We also know that other species of bacteria upregulate expression of an enzyme, lactonase, that degrades AHL removing it from the system. Thus our confidence in this parameter value is fairly low, and we know that it has the potential, via lactonase expression, to be controlled.

Exercise 4.43

(a) What effects would altering the value of γ_A have on the steady-state concentration of AHL and consequently of the steady-state fraction of luminescent bacteria?

(b) Reproduce Fig. 4.22.

(c) Could increasing the rate of AHL removal potentially play a role in "turning off the lights"?

(d) A careful examination of Fig. 4.22 suggests that there is a switch-like dependence of both steady states but the relationship is not identical. Does this make sense given the original data (Fig. 4.17)?

Finally, one of the parameters that can vary among cells is the concentration of the receptor LuxR (R_0). LuxR expression is sensitive to at least two transcription regulator systems and, in some quorum sensing bacteria, may also respond to the AHL-LuxR complex by doubling the expression rate of LuxR. Thus LuxR is definitely worth examining from a biological perspective.

Exercise 4.44

(a) Examine Fig. 4.23 to see consequences of lowering the cell LuxR concentration on the steady-state values of AHL and luminescent bacteria. What is the minimum concentration of LuxR required for the quorum phenomenon to work?

(b) Examine the system of model equations again. Which other parameters might contribute to the threshold value of R_0?

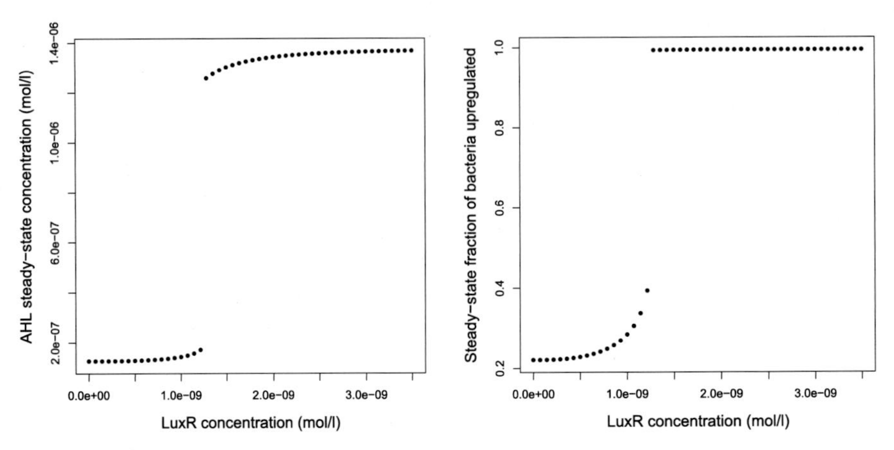

FIGURE 4.23

Steady-state values as a function of LuxR concentration.

The temporal response

By now we should understand the quorum sensing system quite well. However, for our bacteria *Aliivibrio fischeri* and its luminescent cousins, timing is also critical; timing is definitely critical for the host our bobtail squid. Recall that the squid ejects most of its bacteria every morning so that the colony density in each light organ is well below a quorum. However, by evening, about 12 hours later, the bacteria need to be luminescent if they are to be any use to the squid.

Exercise 4.45

(a) Examine the plot you produced in Exercise 4.39(e). Given the parameters we used, about how many hours after starting to grow does the fraction of luminescent bacteria reach its steady state or maximal value?

(b) Take a minute to brainstorm about how this response could happen sooner. Choose three parameters that you think will speed up this process and state why. A look at the model equations might help your reasoning.

(c) Check your predictions in part (b) by creating sliders for your three parameters and plotting the fraction of bacteria upregulated. Explore the effect of these parameters on the time at which the fraction of luminescent bacteria reaches its steady state. Discuss your conclusions.

(d) Then there is the problem of how quickly luminescence will decrease after AHL concentrations drop. The half-life of the luciferase enzymes is about an hour. What does this mean for the behavior of our system? What will happen to the light intensity in the light organs if 95% of the bacteria are ejected?

4.6 Wet lab: hormones and homeostasis—keeping blood glucose concentrations stable

Goals: In this lab, we will study one of the powerful homeostatic mechanisms, the maintenance of blood glucose concentrations after ingesting a large amount of sugar. We will obtain blood glucose concentration measurements and use a simple model of the glucose–insulin system to predict both the insulin response and the rate at which glucose enters the bloodstream following ingestion. The model will also reveal ways to change the experimental protocol to test other aspects of the glucose regulatory system.

4.6.1 Overview of activities

One of the most critical physiological parameters is the concentration of glucose in the blood. Glucose is essential for brain function and is the energy source for many other tissues. However, when its concentration is too high, tissue damage results and significant amounts of glucose are lost via the urine. Two hormones, insulin and glucagon, are responsible for maintaining glucose homeostasis. In this lab, we will learn about the sources and sinks for glucose and the systems that regulate its concentration. We will explore a model with feedback between glucose and insulin concentrations and propose various ways of representing the rate at which glucose enters the blood after consumption (recall caffeine uptake from the gut to the body in Unit 2). After exploring this model for the glucose–insulin system as a function of glucose intake, we will measure our own blood glucose levels before and after ingesting a known amount of sugar to follow the time course of glucose concentration rise and fall. These data will be used to derive model parameters. Once we have basic data on the body's glucose response (extent, time-course), there are multiple additional questions that you may address using thoughtful experimental design.

4.6.2 Introduction to blood glucose regulation and its importance

All living systems control key physiological parameters within tight boundaries. For example, most vertebrates control body temperature, ion concentrations, total osmolality, total volume, blood pressure and, the subject of this lab, the concentration of glucose in the blood. Control implies a fairly sophisticated system including a sensor, a comparison system, and a means of communicating with one or more effectors (e.g., via the nervous and/or endocrine systems) to alter processes in a way that will bring the parameter back into the normal range, thus maintaining homeostasis.

Glucose is an energy source absolutely required by the brain and utilized by all cells in the catabolic pathways that generate ATP, the molecule used to couple energy to biological processes. Glucose is delivered to tissues via the blood, and thus maintaining circulating concentrations of glucose at sufficient levels is essential. Hypoglycemia is acutely dangerous as it causes loss of brain function related to direct and downstream effects of low glucose delivery rate. If glucose concentrations get too

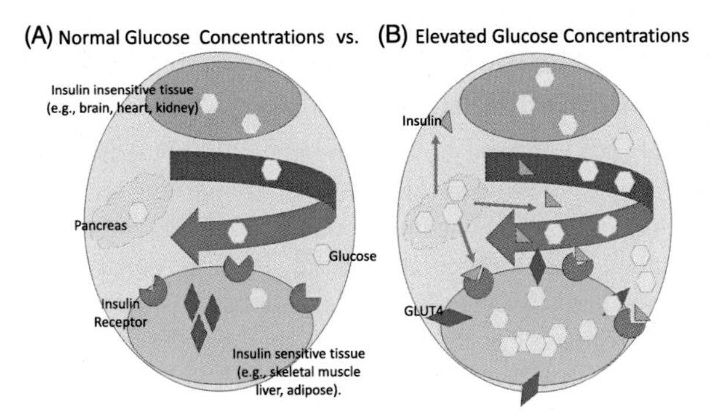

FIGURE 4.24

A comparison of glucose concentrations and tissue glucose uptake when blood glucose is in the normal range (A) versus when it is at a high enough concentration to activate insulin release from the pancreas (B). The tissues are the blood (red arrow), the pancreas (yellow) insulin-sensitive tissues (green) and insulin insensitive tissues (gray). Glucose is represented by the hexagons and insulin by orange triangles that bind to insulin receptors (orange). The result is adding special glucose transporters (blue) to facilitate insulin uptake. Note that the insulin response is to insert new glucose transporters in only some tissues.

high, then damage occurs slowly over time. Hyperglycemia has direct effects including glycosylation of proteins, increasing oxidative damage, decreasing nitric oxide production, and increasing the inflammatory response, all of which damage tissues, especially capillaries and peripheral neurons. Thus uncontrolled hyperglycemia leads to neuropathy, cataracts, and renal failure. The hallmark of uncontrolled diabetes is hyperglycemia.

Fortunately, the natural glucose regulatory system tightly controls circulating glucose concentrations via two hormones, glucagon secreted in response to low glucose and insulin secreted when glucose concentrations increase, for example, after eating carbohydrates and sugars. Lack of insulin (Type I Diabetes) or lack of insulin receptors (Type II Diabetes) both prevent the restoration of glucose concentrations back to normal after a meal resulting in hyperglycemia. These two hormones, glucagon and insulin, maintain homeostasis on a minute time scale to ensure timely responses to rapid increases in glucose use (starting to run) and rapid increases in glucose entry into the body (eating a candy bar). Glucose is used by all tissues of the body and is constantly being removed from the blood. The removal rate depends on the metabolic activity of the tissue; in most mammals, the brain and kidneys are responsible for 60% of the glucose uptake when the body is at rest. When active, the skeletal muscle glucose demand rate increases, and thus the glucose removal rate increases. Glucagon (and adrenaline or epinephrine) act quickly to mobilize glucose from glycogen (a large glucose polymer) and activate metabolic pathways to synthesize glucose to maintain the blood concentration near 5 mM or about 100 mg/dL.

New glucose supplies arrive via the episodic ingestion of food. Foods are digested into small molecules such as glucose in the small intestine and transported into the body increasing their concentrations in the blood and total extracellular fluid. Even small increases in glucose concentration are detected by the pancreatic β-cells; increased glucose entry initiates a cascade of events to release insulin. Insulin circulates through the body and binds to receptors on its target tissues i.e., skeletal muscle, adipose tissue, and liver (see Fig. 4.24). From stimulus to effect takes 5 to 15 minutes. These insulin sensitive tissues respond by inserting special glucose transporters into their cell membranes, thereby greatly increasing their capacity for glucose uptake, essentially increasing the V_{max} for glucose removal from the blood. Once in these cells, the glucose is either oxidized for ATP production, converted to other metabolites or converted to glycogen for storage. Insulin favors the latter by increasing glycogen synthesis and inhibiting its breakdown. Insulin (and glucagon) are removed continuously by several degradation processes. (See a physiology text such as Sherwood et al. [63] for a more in-depth overview).

Exercise 4.46

Using the information in the previous paragraph, draw a diagram indicating the sources and sinks of glucose and how these are affected by insulin and glucagon. Indicate what effectors increase or inhibit the release of these two hormones and how they are removed from the circulation.

Normal fasting glucose concentrations are 3.5–5.5 mM (63–100 mg/dL). Glucagon secretion from pancreatic α-cells increases as glucose levels decrease. Low levels of insulin are secreted even when glucose is as low as 3 mM, but the rate increases steeply as glucose concentrations rise. In this lab, we will focus on the glucose–insulin system. Our goal will be to understand the quantitative and temporal relationships between them. Models of their interactions are critical for the maintenance of blood glucose in diabetes Type I patients for whom insulin has to be given exogenously in exactly the right amounts at the right time to prevent excessive hyperglycemia without inducing a coma due to inadvertent hypoglycemia.

It is always good to pay attention to units. Physiological concentrations are generally reported in moles per liter or in this case mM, but most clinical measurements are reported in grams per 100 mL. 100 mL is 1 dL (a tenth of a liter). The molecular weight of glucose is 180 g mol^{-1}, so you can easily convert between these values. Both are used in the literature. Your glucometer will read in mg/dL.

Exercise 4.47

Use the information given to draw the graph of $[\text{glucose}]_{\text{blood}}$ vs. time that you would expect if you ate a candy bar. Add the expected plot for insulin on the same graph. Write an explanation of the graph you drew.

4.6.3 Developing a model

Here we will explore a system of equations to model the relationship between glucose and insulin after ingestion of glucose. Let G (in mg/dL) and I (in ng/dL) represent the

Table 4.4 Parameter values in the glucose model in Eq. (4.34).

Parameter	Value	Units
G_{rel}	1.5	
k	4×10^{-4}	
r_1	0.005	
v	30	
G_m	120	
r_2	0.2	

plasma concentrations of glucose and insulin, respectively, and let t represent time in minutes. The change in plasma glucose and insulin concentrations can be described by

$$\begin{aligned}\frac{dG}{dt} &= (\textit{release}) - (\textit{uptake}) + (\textit{ingestion}),\\ \frac{dI}{dt} &= (\textit{secretion}) - (\textit{degradation}).\end{aligned} \tag{4.33}$$

Exercise 4.48

(a) In Eq. (4.33), discuss biologically why each term is included.

(b) Sketch a two-compartment qualitative diagram of this system and include dependence links to show what variables will likely be in each term (e.g., if the flow will depend explicitly on G and/or I) and discuss why you included each dependency.

There are various ways to mathematically represent each of these terms. Here, we propose the simple model

$$\begin{aligned}\frac{dG}{dt} &= G_{\text{rel}} - kGI - r_1 G + \frac{G_{\text{ing}}(t)}{V},\\ \frac{dI}{dt} &= \frac{vG^4}{G_m^4 + G^4} - r_2 I,\end{aligned} \tag{4.34}$$

with approximate values of the parameters given in Table 4.4. We will investigate whether or not this simple model can capture the general dynamics of a glucose spike following ingestion of sugar.

Exercise 4.49

Work through the exercises below to help you better understand the model.

(a) Given that G is in mg/dL, I is in ng/dL, and t is in minutes, fill in the units of each parameter in Table 4.4.

(b) Although the uptake term could be split into various rates of uptake by the brain, red blood cells, peripherals, kidneys, and liver, here we combine these into insulin sensitive and not insulin sensitive with the expression $f(I)G$ where $f(I) = kI + r_1$. Sketch $f(I)$ versus I and discuss the assumptions made by this representation.

(c) The insulin secretion term is given by the Hill function $g(G) = \frac{G^4}{G_m^4 + G^4}$ multiplied by the maximum secretion rate v. One study [25] reported that the insulin secretion rate reached half its maximum value at 120 mg/dl glucose (which corresponds to G_m) and reached its maximum at around 280 mg/dl glucose. Compare these values to what is assumed in $g(G)$ by plotting $g(G)$ versus G. The Hill coefficient is equal to 4, but what happens if you change this? Compare the form of this insulin secretion term to the Michaelis–Menten type of model or a Holling Type III functional response (see Unit 3).

(d) The insulin degradation rate term is given by $r_2 I$. What is the physical interpretation of r_2?

(e) The parameter V in the model is the volume of extracellular fluid, which is equal to 150 dL in our typical 70 kg man (see Unit 2, Pharmacokinetics Case). What are the units of the ingestion term $G_{\text{ing}}(t)$? Suppose you ingest a candy bar that contains 30,000 mg of simple sugars, and assume that it is completely absorbed in 30 minutes. The simplest representation of $G_{\text{ing}}(t)$ would be a step function given by

$$G_{\text{ing}}(t) = \begin{cases} h_1, & 0 \le t \le 30, \\ 0, & t > 30. \end{cases} \tag{4.35}$$

A sketch of this function is given in Fig. 4.25. Assuming that 30,000 mg of sugars are ingested, what is h_1?

(f) In part (e), we assumed that the rate at which glucose enters the bloodstream is constant. Sketch other possible forms of $G_{\text{ing}}(t)$. What is the logic behind your sketches?

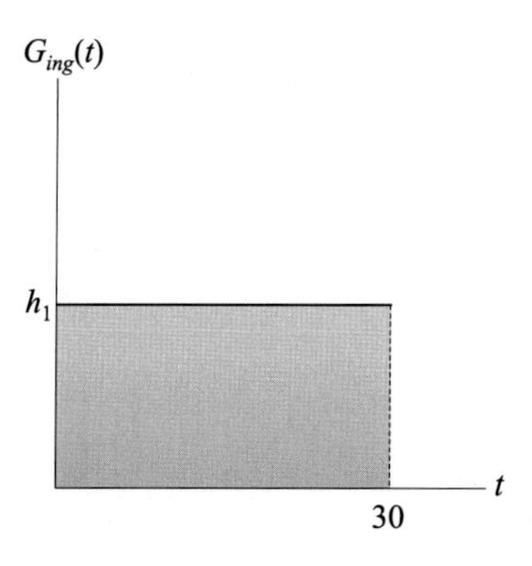

FIGURE 4.25

Constant function for the glucose ingestion rate.

Exercise 4.50

We will explore this model in stages beginning with a fasting situation, that is, $G_{ing} = 0$. Input the code in Listing 4.19 into R (or open up the file unit3ex19.R). This file solves Eq. (4.34) for $G_{\text{ing}} = 0$ and plots the numerical solution. Insert comments in the script to explain each line.

Listing 4.19 (unit4ex19.R)

```
library(deSolve)

gluc = function(t,y,parameters) {
  with(as.list(c(y,parameters)), {

  G_ing = 0

  G_uptake = (k*I+r1)*G
  I_secretion = v*G^4/(Gm^4+G^4)
  I_degradation = r2*I

  dG = G_ing + G_rel - G_uptake
  dI = I_secretion - I_degradation
  list(c(dG, dI))
})}

t = seq(0,100,length.out=100)
parameters = c(G_rel=1.5,k=4e-4,r1=0.005,Gm=120,v=30,r2=0.2)
yinit = c(G=100,I=25)
out = ode(y = yinit, times = t, func = gluc, parms=parameters)
plot(out[,1],out[,2],type="l",
     xlab="time (minutes)",ylab="plasma concentration",ylim=c(0,150))
lines(out[,1],out[,3],col="blue")
legend("topright",c("glucose (mg/dl)","insulin (ng/dl)"),lty=c(1,1),
       col=c("black","blue"))
```

Exercise 4.51

Homeostatic processes ensure steady-state internal conditions; here we explore how glucose homeostasis is maintained.

(a) What is the homeostatic concentration of glucose assumed by this model? What is the homeostatic concentration of insulin? (Note that the homeostatic level is the same as the steady state of the model.) Do not forget to note the units.

(b) If you simulate the dynamics following ingested glucose, you should see that the concentrations of glucose and insulin will rise and then return to their homeostatic levels. Assume that $G_{ing}(t) = 450$ (mg/min) for $0 \leq t < 30$ and $G_{ing}(t) = 0$ for $t \geq 30$. You can do this by using an if-else statement. Inside the function gluc, replace `Ging = 0` with the following:

```
if (t<30)
     G_ing=450
else
     G_ing=0
```

Plot $G(t)$ and describe the observed dynamics.

(c) The function $G_{ing}(t)$ proposed in part (b) assumes that the total amount of glucose eaten is taken up into the body. Calculate this amount in mg recalling that the units for $G_{ing}(t)$ are mg/min.

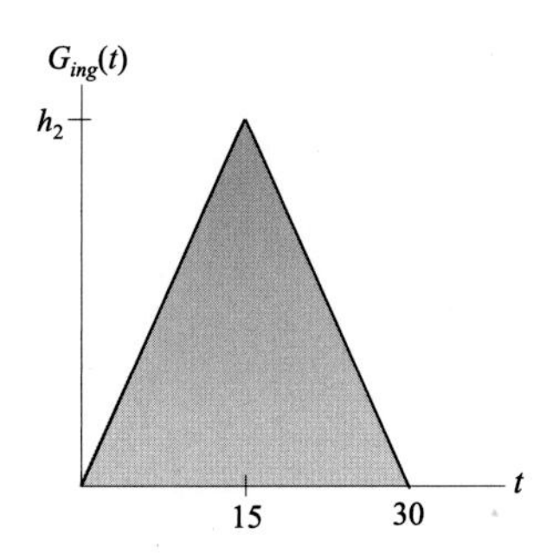

FIGURE 4.26

A triangular function or hat function for the glucose ingestion rate.

Exercise 4.52

Eating a candy bar will release a surge of glucose into your bloodstream and give you what is often called a "sugar high." A typical candy bar contains about 30 g (30,000 mg) of simple sugars, and the glucose from the bar might be absorbed in 20 to 30 minutes. Below we will simulate the glucose dynamics following ingestion of the candy bar.

(a) To be realistic, we need to choose a glucose ingestion rate function (e.g., step or triangular), $G_{\text{ing}}(t)$, that will cause the glucose concentration to spike. Set up the code for both a constant ingestion rate (see Fig. 4.25) and a triangular function (see Fig. 4.26). Make sure that both rate functions assume that 30,000 mg of glucose is ingested over 30 minutes. Set up the explicit calculations for G_{ing} in your code.

(b) First, to separate the effects of glucose ingestion from the insulin response, set up the model without the insulin response curve, that is, $G' = G_{\text{rel}} - r_1 G + G_{\text{ing}}(t)/V$. In R, solve this for $G(t)$ and plot the concentrations of blood glucose using a) the step model and b) the triangle model for $G_{\text{ing}}(t)$. What do you observe? When is the blood concentration of glucose maximal, and what is the maximum value? Does this make sense? Why? Is this a realistic model of diabetes? Why or why not?

(c) Now run the simulation with the insulin terms back in the model (e.g., the insulin-dependent glucose uptake term and the glucose-dependent insulin equation) using both the step and hat functions for G_{ing}. Plot the respective glucose and insulin curves. How do the responses differ for the proposed ingestion rate functions? How do the glucose curves differ from those plotted without an insulin component. What feature(s) of the model explain the different shape and magnitude of the blood glucose concentration trajectory?

(d) Note when the glucose concentration begins to rise, when it peaks, and when it returns to normal. How long is the lag for insulin secretion rate to increase? Why does insulin secretion lag behind the rise in blood [glucose]?

(e) Why does the insulin level decrease after a certain amount of time?

(f) If you eat a candy bar at $t = 0$ and another one at $t = 20$, then how do you think your blood glucose and insulin levels would respond? Run the simulation. How does the model output compare with your predictions?

(g) Foods like beans, rice, oats, fleshy fruit, vegetables, whole-grain bread, and pasta contain their glucose in a large polymer called starch, which requires an extra digestion step to break the starch into glucose. The breakdown rate is slower than the absorption rate for simple sugars, that is, the digestion step is rate limiting for the uptake of sugars from starch. Here we will simulate the response of a body to eating pasta, a classic starchy food. Choose a glucose ingestion rate of about 300 mg/min maintained for 120 minutes. What is the model output? How do the blood glucose and insulin curves differ from the same graph generated by eating a candy bar? Why are they different?

4.6.4 Experiment: measuring blood glucose concentrations following glucose ingestion

Introduction to the procedure

We will study the time course of blood glucose concentration in response to ingesting about 30 g of glucose and model the insulin response. Other important hormones such as glucagon or pathways may be explored in our follow-up exercises or as research projects. We will measure blood glucose concentration in *volunteer* subjects (see below) with a glucometer before and after sugar ingestion. Blood will be taken by finger-prick every 10–15 minutes, and glucose concentration measured and recorded. These are the input data for the model and will reflect the rate of glucose absorption from the small intestine and the effect of any subsequent insulin response.

NOTE: This experiment is on human subjects and REQUIRES EVERYONE to take appropriate precautions with the samples and with the information we gather.

1. Subjects (those who will provide blood samples) must read and sign a consent form before participating in this exercise. NO ONE may be coerced into taking a series of blood samples. Subjects should realize that your blood glucose concentrations will be revealed to others in the class. NO ONE should reveal his or her health status or reasons for opting out. Any subject may opt out at any time.
2. ALL participants, subjects, recorders, and observers are to follow the safety rules regarding the lancets and glucometer test strips. No one is to share the names of the subjects nor their personal data beyond the confines of this class. Health information is **never** shared without explicit permission.

We will measure glucose concentration in mg/dL of plasma, the standard units used by health care professionals. We will assume that insulin release occurs as predicted by the model, that is, insulin concentration rises steeply in response to rising glucose concentrations. Insulin measurements are expensive, but fortunately mathematical models allow us to project the insulin response from the [glucose] data.

In preparation, review the conceptual model of the glucose–insulin–tissue system by updating the drawing you made in Exercise 4.16. Name some variables that might alter the extent or time course of glucose rise and the insulin response and explain how each variable might work. Sketch a graph showing your predictions.

The data will be most interesting if subjects come to lab in a relatively fasted state, for example, nothing to eat or drink besides water for an hour or two beforehand. Consider what might work best given your needs and schedule. Plan accordingly.

Experimental procedure

Subjects and assistants will be identified. *If anyone wishes to opt out of being present for data collection, please let one of the instructors know.*

Materials:

1. Glucometers, lancets, lancet holders, and glucose test strips. Each subject will take up to 10 measurements.
2. Sterilization and first aid supplies: alcohol wipes, gauze wipes, BandAids.
3. Biosafety supplies: gloves, test strip disposal bag, sharps disposal container, and disposable absorbent paper for the lab bench.
4. Sugar source provided outside the lab for safety. Each subject will ingest 30 to 50 g of readily absorbed sugar as jelly beans or orange juice. Use the sugar concentration per serving as listed on the package to calculate how much the subject needs to eat or drink to ingest about 50–60 g of sucrose; sucrose is a dimer of glucose and fructose.

Prior to collecting data, become familiar with your equipment. Read the instructions carefully and look at each piece. Prepare the lancet in its holder. Make sure there are test strips in the glucometer. If you are assisting someone else, please put on gloves. Subjects may wish to note when and what they ate most recently. Recall that ideally each subject has fasted at least two hours. Once subjects and assistants are ready, take a baseline measurement:

1. Make sure the test strip is in position in the glucometer ready for use. Read the directions for the particular glucometer you are using!
2. Clean the finger to be poked with an alcohol wipe. Place the lancet device firmly on the tip of the subject's finger slightly off to the side. Press the button that will push the lancet into the skin. A blood drop will form. Wait a few seconds for the drop to get larger; massaging or gently squeezing the finger may help generate a large enough drop of blood.
3. Bring glucometer and test strip up to and below the drop of blood. The blood will be drawn into the strip by capillary action. Hold the strip against the blood drop until the glucometer signals that it has received the sample. The results will be displayed soon thereafter. Confirm that your instrument is reading in mg/dL. Record your result.
4. Sometimes the reading is not successful, and there is an error message. Two likely reasons for this are that there was insufficient blood or the instrument battery is low. Consult with your instructor and retake the sample.
5. Recap the used lancet and dispose of it in the sharps container.
6. Eject the used test strip directly into the test-strip disposal bag by pressing the release button on the glucometer.

Once you are satisfied with your baseline glucose concentration measurement and have reset your meter for the next measurement, it is time for subjects to ingest a known amount glucose. Wash your hands and then as quickly as you comfortably

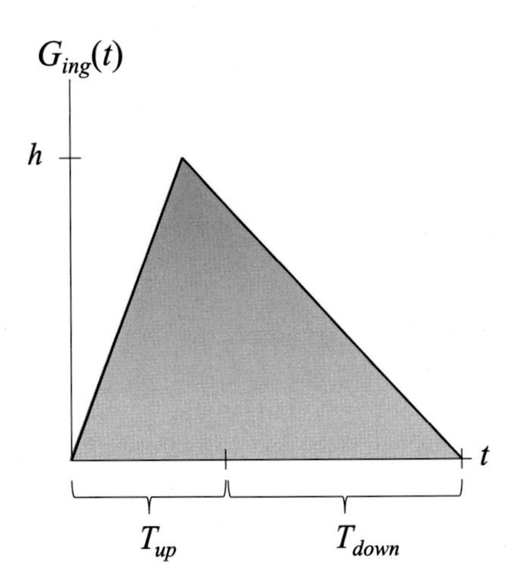

FIGURE 4.27

Proposed function for the glucose ingestion rate is triangular with a faster initial ingestion rate.

can, ingest the requisite amount of orange juice, jelly beans, or other sugar source provided. (You may wish to take in less than planned.) Record the amount you did ingest and calculate how many mg of glucose you took in.

Repeat glucose measurements at 10–15 min intervals for the rest of the lab period; after the initial rise in glucose, you may wish to wait 20 min between samples. Record and plot the data as you go along.

4.6.5 Analysis

You will need the packages `manipulate`, `deSolve`, and `FME` for the analysis. Make sure these are installed and loaded, and then input your data directly into R. We will fit the differential equation model proposed in Eq. (4.34) to observed blood glucose concentrations to obtain estimates of the ingestion rate.

We will compare our data to those predicted by using the parameter estimates (Table 4.4) and an estimation of the subject's extracellular fluid (ECF) volume. The ingestion rate G_{ing} will be fit assuming a triangular function as explored in Exercise 4.51 and shown in Fig. 4.27.

To estimate the subject (ECF) volume (i.e., the volume to which the glucose will disperse) use the relationship that the ECF volume $(L) = 0.20 \times$ body weight (kg) (recall that 1 kg $= 2.2046$ lbs) (Fig. 2.10). Just for fun, estimate the maximum concentration of glucose you would see in the absence of insulin given the amounts ingested and the ECF volumes of the subjects in your research group. Recall that in a clinical situation like this one, volumes are expressed in dL.

Exercise 4.53

Follow the steps below to formulate the ingestion function, calibrate the model, and compare the model output to the data.

(a) Use the amount of glucose actually consumed in milligrams, that is, the *dose*, to determine the area under the curve $G_{ing}(t)$. Show this calculation in your code.
(b) Note that there are three unknowns, the times during which the absorption rate G_{ing} was increasing and then decreasing, T_{up} and T_{down}, respectively, and h in the function $G_{ing}(t)$, shown in Fig. 4.27. Assume that the ingestion function is symmetrical, so that $T_{down} = T_{up}$. Write h in terms of T_{up} and known *dose*. Now there is only one unknown parameter T_{up}, which we will obtain by fitting the data to the model. Write $G_{ing}(t)$ as a piecewise function in terms of T_{up} and *dose*. The variable we will manipulate is T_{up} to attain a reasonable match between the calculated and measured blood glucose concentrations.
(c) Input your $G_{ing}(t)$ expressions into the function `gluc` from Listing 4.19. Choose a value of T_{up} and run your code to make sure it works!
(d) Recall that the optimization algorithm requires an initial guess for T_{up}. Create a slider for T_{up} to obtain an estimate for this parameter. In the `manipulate` function, make sure that you plot both the model output and the data.
(e) Create a cost function that calculates the squared residuals between the model output and the data. Then use `modFit` to find the best-fit parameter values. Record these values. *Refer to Listings 4.6 and 4.8 for an example of writing a cost function and using* `modFit`.
(f) On the same graph, plot the data and the model output using your best-fit parameters.
(g) How well does the model fit your measured glucose concentrations? Does the behavior of insulin make sense? Explain.

Exercise 4.54

Now that you have calibrated your model, consider the following questions:

(a) What assumptions were made in the function $G_{ing}(t)$? Try using the function in Fig. 4.27 with two unknown parameters that you must calibrate, T_{up} and T_{down}. Show explicitly how you will express these two pieces of $G_{ing}(t)$ using the parameters you know. Does this function fit the data better? Does it make more sense biologically?
(b) Instead of a triangular function, what other forms of $G_{ing}(t)$ might you suggest? Think back to the pharmacokinetics case study in Unit 2.
(c) Would this triangular form of $G_{ing}(t)$ be a good choice for modeling the ingestion of a pasta meal? Why?
(d) How might you quantify the difference between the model and output? One way of quantifying the difference is calculating the sum of the squared residuals. Are there other ways to quantify how well your model fits the data? For example, consider other characteristics of the output, such as the breadth of response (e.g., width in time at half-maximum value of glucose).
(e) Suggest an experiment or two that would test this model clearly indicating which aspect of the model would change and why.
(f) Consider how you might perform a sensitivity analysis for this model. Why might this be an important step in the modeling process?
(g) How might you modify the model to consider other variables such as increased glucose use rates (exercise) or anticipatory increases in the regular release of glucose from cellular stores?

4.6.6 Thoughts to consider for potential follow-up experiments

Revisit the model you drew describing the dynamics of glucose regulation. Among the aspects that we ignored are the roles of circadian rhythms and the central nervous system in stimulating anticipatory and food-related insulin release [30,43]. When the CNS expects a large ingestion of sugar, neurons from the autonomic nervous system stimulate β cells directly to release insulin. Expectation might be conveyed by the sight of, or odor from, sugar-rich foods (cinnamon rolls, carrot cake) or by stimulation of the tongue sugar receptors. If that food is not ingested, then what will happen? How might you test this idea?

Does the type of sugar or food alter the amount or the rate of glucose uptake from the small intestine? In our prior experiment, we ingested sucrose, a disaccharide that is rapidly broken into glucose and fructose rapidly delivered to the circulation via the small intestinal epithelial cells. Fructose is handled differently from glucose by the body, and therefore the type of sugar might make a difference. For example, would ingesting primarily fructose (e.g., agave syrup) alter blood glucose responses? Why? What would happen if you ingested the monosaccharide glucose instead of the disaccharide sucrose, or maltose, a disaccharide made of two glucose molecules vs. sucrose, which is a disaccharide of glucose and fructose? Starch, a large glucose polymer in breads, potatoes, and pasta, must first be digested by amylase enzymes to glucose molecules, which can be transferred to the blood. Consider how the source of sugar will alter the magnitude and temporal change in blood glucose concentration? What factors would you need to consider when designing an experiment to test your ideas?

Another key question is "how long does insulin remain in circulation?" We can use our model and the time constant for insulin disappearance to estimate the decay from peak levels, but can we estimate the time window of effective insulin response experimentally? Think about how we could design an experiment using our only tool, that is, measuring blood glucose, to determine how long insulin concentrations or the insulin response (extra GLUT4 transporters in muscle cells) persists.

Finally, revisit the model in Eq. (4.34). What other factors might alter blood glucose concentrations directly or indirectly? How might we access these experimentally? If glucose concentrations go down, for example, due to increased use by active muscles, what term(s) would we want to add to the model to ensure homeostasis?

If there is time for a follow-up lab, then prepare to test one of these ideas or one of your choice. Analyze the data using the model as is or modified appropriately for your experimental design. Compare your new results to those obtained previously in terms of ingestion time estimates, the height of the glucose spike, the width of the spike at half-height, or the insulin response time. Explain in terms of the biological manipulation and response.

A Final Word – especially to students

The overarching goal of all of the exercises in this book is that you will feel more confident in approaching messy complex problems and organizing your thinking through mathematical models. Although iterating between data collection and formulating and analyzing models is rarely smooth, it is usually rewarding as new insights are revealed with the only cost being the time you engage with an interesting problem.

Now you are in a position to carry out your own projects and address questions that are most interesting to you through models and/or experiments. The case studies and labs in this text all lead to follow-up questions. What analysis, modifications to the model, or experiments could you design to help you answer those questions? Moreover, throughout this book, we typically did not test the robustness of the models; after calibrating a model, can you design new experiments to test whether or not your model can accurately reproduce new observed data?

We hope the approaches used in this book will be helpful in future classes and beyond. We also hope that this introductory text stimulated your interest to explore some of the modeling tools more in-depth. A breadth of tools, including statistical modeling, difference and differential equations, computer programming, and experimental design, are required in modeling biological systems. All these tools can be explored further in various undergraduate courses.

UNIT

Technical notes for laboratory activities

5

5.1 Introduction

All of the wet lab activities presented in this text are suitable for students with little or no lab experience. They are flexible, that is, can be done in multiple ways depending on the facilities, personnel, and goals of the course. The following technical notes include comments on the data that are ideal to gather for the lab, practical tips on how to do the lab as outlined in the text, suggestions for other systems that would work well, some dry-lab and even some *in silico* lab experiences that might be beneficial. For mathematicians, we recommend finding a biology partner or consultant who can help you navigate your institution's equipment, what might be done in a classroom setting or substitute labs effectively.

We do these labs in a tight two-hour time frame (three would feel ideal) by completing the prelab work as homework first, frequently devoting a lab period to preparation. Often, data analysis is completed during our "lab time," especially when data sets need to be combined or shared.

If the class has a mix of natural and physical science majors along with mathematics and computer science students, then the laboratory opportunities are a good venue for reinforcing teamwork skills with those with more lab experience mentoring those with less and vice versa when it comes to data manipulation in R. The first two labs as written in the text require pipetting and measuring absorbance. It can be helpful to do the "Prelab" lab (in Unit 1) first, especially if most of the students are lab novices; the intention is to introduce basic equipment that will be used subsequently along with the idea of recording, graphing, and interpreting data. The Prelab Lab also fits in nicely with a lab safety program concerning labeling, liquid handling, spills, and waste disposal. Note that like all the labs, this is intended only as a prompt to help you think about what students might need to be successful.

5.2 Population growth

There are many ways to witness and measure population growth in the lab. We present a fairly classic bacterial culture lab in the text, but it is easy to substitute common baker's yeast or, when the time frame is longer (e.g., quick checks every day for a week or so), phytoplankton growth (see section on "alternate organisms"). Anything

Exploring Mathematical Modeling in Biology Through Case Studies and Experimental Activities
https://doi.org/10.1016/B978-0-12-819595-6.00011-6

that quickly reproduces would work (ciliate cultures, aphids). As written, the goal is to observe logistic growth by measuring the population as it transitions from exponential to carrying capacity-limited growth. We use difference equations and plot the per capita growth rate at each time point vs. the population size to obtain estimates for the growth rate r and carrying capacity K. We vary initial population size to ask whether the value for r or K depends on population density (r does, but K does not). In a follow-up lab, we vary nutrient concentration and ask again if the parameters r and K change. The follow-up or variations on this lab can give students tools to ask additional questions about population growth in independent projects.

Bacterial growth media and tips

This lab requires preparation well ahead of time to get the bacterial cultures ready and to prepare and autoclave media and glassware.

When growing bacteria for student use, it is important to have a "safe" strain such as common lab strains of *Eschericia coli*. Bacterial cultures are typically stored on streak-plates, so that it is necessary to rejuvenate them before class. Two days before, use a sterile swab or loop to inoculate two tubes (although one is enough, we always prepare two to have a back-up) containing 5–10 mL of nutrient broth with *E. coli* and grow overnight at 37° while shaking to aerate (at about 120 rpm). The next day, innoculate one or two 250 mL flasks with 50 mL nutrient broth with about 1 mL of this culture and let it grow overnight. This is the source "overnight" culture for the class. About one hour before lab, inoculate the flasks that students will use for their cultures: provide one per group and vary the initial amount of bacteria, for example, 0.25, 1.0, and 4 mL per flask (the starting densities will be on the order of 10^7 bacteria/mL). The bacteria should be actively growing by the time students start. They should take initial absorbance measurements at 600 nm as soon as possible using pure nutrient broth as a blank. Repeat measurements at 20 or 30 min intervals for up to 8 hours. Remind students to return their flasks to the 37° shaking incubator as soon as possible after taking the aliquot to measure: growth slows significantly without oxygen and at lower temperatures!

Nutrient broth or other suitable media such as LB broth is prepared according to the package directions by thoroughly dissolving the powder in deionized water. It is dispensed in flasks (25 mL in 125 mL flasks or 50 mL in 250 mL flask) and capped tubes as needed prior to autoclaving for 15–20 min. The sterile media can sit for a week or more on the bench top as long as each flask is covered with foil or capped. If you are going to do the optional plate-count portion of the lab, then nine nutrient agar plates are needed per lab group. Sterile plastic 100 mm plates require about 25 mL of nutrient agar each. Prepare the nutrient agar in bulk, autoclave, cool somewhat, and then pour while still warm. This can be done up to a week or so ahead of time if the plates are stored sealed in plastic in the refrigerator.

Other items need to be autoclaved as well. Each lab group will need access to racks of sterile pipet tips for 1 mL and 100 μL measurements. If you are doing plate counts, then each group will need sterile small tubes (0.75–2.0 mL) and spreaders (disposable plastic ones are available, or you may wish to flame sterilize glass ones).

The polystyrene semimicrocuvettes need not be sterile. We reuse them by dumping them and their contents into a dishpan of diluted bleach to sit overnight. They can be rinsed and air dried by the next day for reuse.

Make sure to provide table-top waste containers for tips and used spreaders and tubes. These can be combined and the waste autoclaved in autoclave bags for safe disposal.

If you have access to a plate reader instead of spectrophotometers, then have students set up their experiments (several groups per plate) in sterile 48 or 96 well plates adjusting the volumes as needed per well (e.g., 200 μL for 96-well plates). This is a great opportunity for testing more initial conditions! Do not forget to have a "no bacteria" blank for each group. Set up the plate reader to collect data at 10-min intervals at 600 nm for 12 hours (or more if you like) while shaking and holding the samples at 37° ideally with a lid (only if your plate reader can warm the lid to prevent condensation). The advantages of the plate reader are that significantly less media and plastic-ware are needed, the data are collected automatically, and there is no need to access the lab after official lab hours. One disadvantage is that the students do not directly "see" the growth unless you keep the source flasks growing in the incubator and have students inspect them periodically.

One follow-up lab suggestion is to vary the amount of energy available by preparing the media at 1/4 to 4 × the standard concentration. There are an infinite number of ways to manipulate the type of media or add glucose or sucrose; these variations make good student projects.

Optional plate counts and converting optical density to numbers of bacteria

There are two reasons to perform plate counts. One is to determine the number of viable bacteria vs. particles adding to the optical density measurement. The second is to determine the relationship between the value of Abs_{600} and the density of bacterial cells in cells/mL. There are published values that can be used: in our hands, Abs_{600} of 1 is about 8×10^8 cells/mL similar to published values. There is a caveat: as scattering increases, the absorbance value underestimates the number of cells, that is, it is sublinear, so that using this conversion will overestimate the initial growth. Therefore as absorbance is greater than about 0.5, it is ideal to dilute the solution in nutrient broth to get an accurate reading. Most people do not do this, and for our purposes, it is OK since we are applying the same conversion to all cultures.

Follow the directions given in the text for preparing bacterial samples and plating them on the agar plates. For the early time points, there will be about 10^7 cells/mL, and you will expect about 100 colonies on a plate from 100 μL of the 10^{-4} dilution. As the bacteria grow to near 10^9 cells/mL, there will be too many colonies to count at dilutions of less than 10^{-6}. Be sure to have students prepare plates at different times during the growth period to get a good curve. Once the bacterial samples are spread, seal the plates and incubate overnight or longer at 37° until colonies are visible. Although the work is tedious and many of the plates will be unusable, students do enjoy "seeing" their 10-fold dilutions in the plate counts.

An alternative approach to forming the standard curve relating OD to the density of bacteria is to start with a high concentration of cells, from which each student has to prepare a series of dilutions and plates for determining the cfu in this high density colony. Then prepare a series of dilutions with sterile media to obtain a series of samples representing 0.75, 0.5, 0.25. 0.1, 0.05, and 0.025 × the original. Measure the OD_{600} for each dilution. Once you have the estimate for the cfu in the original culture, you can calculate the cfu in the dilutions and plot this value vs. OD to obtain the standard curve.

Alternate organisms

Although many biology departments routinely grow bacteria and can support this lab, most mathematics departments do not! Similarly, some biology departments may maintain cultures of ciliates or algae that would be easy to access for this lab.

One of the easiest alternatives is to grow ordinary baker's or brewer's yeast, strains of *Saccharomyces cerevisiae* in dilute apple juice, or other sugary medium. The advantages are in the lower expense and time to prepare. The disadvantage is that it takes even longer for yeast to grow to carrying capacity, up to 30 hours. We have prepared stock concentrations of yeast from dry yeast packets for the class or had students prepare their own cultures. The medium we use is a 1:9 dilution of apple juice:deionized water. To start, 0.1 g of dry yeast is added to 10 mL of the dilute apple juice and gently mixed. Each group will have two 125-mL flasks in which to combine the stock yeast culture and dilute apple juice for a total volume of 10 mL. Recommended ratios of yeast:apple juice 1:39, 1:19, 1:9, 1:4, and 1:1 (e.g., 0.25 mL:9.75 mL, 0.5 mL:9.5 mL, etc.) will give good growth rates, which vary significantly with the initial population size. Yeast can grow on the bench, but ideal growth occurs at 30°C with shaking to aerate (cover the flasks loosely with foil).

Yeast population size can be measured spectrophotometrically at 550 nm using the dilute apple juice as a blank. With 10 mL cultures, it is necessary to return the cuvette contents back to the culture flask each time. Alternatively, increase the size of the flasks and culture volume sufficiently for as many as 20 samples over the 30-hour time course needed to reach carrying capacity (about 5–6 × 10^7 cells/mL). To convert Abs_{550} to numbers of yeast cells, you can assume a linear relationship with an absorbance of 1 equivalent to 3 × 10^7 cells/mL. One caveat is that as absorbance increases beyond 1, the measurement underestimates the number of cells due to multiple scattering and limitations of the instrument itself. Thus it is a good idea to dilute samples with high readings into the apple juice medium (e.g., 1:1, 1:4, or more) to get an accurate absorbance value: do not forget to correct for the dilution step! Alternatively, yeast are small, but big enough to count under the microscope at high power. Ideally, counting slides (hemocytometers) will be used, but it is always possible to count several fields or view, average the counts, and report cells/field of view. It may be necessary to dilute the cultures using the serial dilution techniques described in the lab write-up, to have a small enough number to count, that is, less than 100 per field. A nice addition to using yeast cultures is that simple urinalysis sticks can be used to follow the change in glucose concentration during the growth period, and ethanol hydrometers can be used do determine the accumulation of this anaerobic metabolic waste product.

Unicellular algae can also be grown relatively simply on the bench top but will take a week or more to grow to carrying capacity. *Clamydomonas* species and appropriate growth media (e.g., Bold's basal medium) can be purchased through biological supply companies. Preparing this lab will take time as a uniform stock culture of about 10^6 cells per mL will be needed to inoculate the experimental growth flasks containing medium to about 10^4 cells/mL. The population density should be monitored daily by counting the cells as described before for yeast or by measuring the absorbance at 680 nm (chlorophyll). Although this growth experiment takes a longer time, it avoids late-night sampling and is safe. One caveat, light is needed (300 foot candles) but not too much (a sunny window sill is too much). Yellowish cultures suggest too much light.

5.3 Enzyme kinetics

The goals of this lab are to attain a full enzyme kinetic data set to compare multiple solutions of the Michaelis–Menten equations to estimate the parameters K_M and $V_{\max}$ and subsequently use apparent changes in these parameters to determine whether an inhibitor is competitive or noncompetitive. There are many enzymes that could be used for this lab, and our choice of catechol oxidase (also called polyphenyloxidase) from apples was driven by the opportunity for students to prepare the crude extract themselves, the low expense, and the existence of well-characterized competitive and noncompetitive inhibitors to give more incentive to finding the kinetic parameters K_M and $V_{\max}$.

Tips and solution preparation for catechol oxidase

Catechol oxidase can be extracted from all types of apples except for varieties designated as *Arctic*: these are engineered to not express the browning enzyme [5]. Variations by apple variety and age are to be expected. Bananas, pears, and potatoes are also good sources and may be prepared identically to the apple extracts.

One difficulty with using catechol oxidase from a natural source is that the activity can vary with type of apple, how old the apples are, and how well the cells are opened to release the enzyme. This means that it is really important to preview the enzyme activity at an intermediate concentration of catechol (e.g., 8 mM) to make sure the conditions will work. The amount of enzyme and the duration of the reaction can be adjusted.

We buy several apple varieties, slice them, and watch for browning before inviting each student group to select which one to use to prepare the enzyme; recall that the extent of browning in sliced fruit depends on both the presence of natural substrate and the activity of the enzyme.

Catechol, the substrate, K_M can vary with the source of the enzyme but can be as low as 2–3 mM up to 12–13 mM. For the sake of time and accuracy, we suggest making a stock solution from which the test concentrations are prepared for the students ahead of the lab. It is important that these stay refrigerated and in the dark (dark bottles or in a box) to prevent light-induced degradation of catechol. Catechol has a molecular weight of 110.1 g/mol and is highly soluble in water (up to 4 M), so that concentrated stocks are feasible: we prepare a 28-mM stock in 100-mM phosphate buffer or MOPS buffer at pH 7. We then dilute the stock with additional buffer to end up with final concentrations in the reaction mixture that are easy "round" numbers as shown in the following example table assuming a ratio of 1-mL substrate solution to 0.4 ml of enzyme plus water or inhibitor solutions. Other strategies will definitely work.

Recipes for 20 mL of the substrate catechol							
final [catechol] in the reaction mM	0.5	1	2	4	8	16	20
mL pH 7 buffer	19.5	19	18	16	12	4	0
mL 28 mM catechol	0.5	1	2	4	8	16	20

We have used both spectrophotometers and a plate reader in absorbance mode with automated enzyme injection to initiate the reaction; the latter is handy since each group can prepare the reaction in a 96-well plate so that the entire kinetic curve, a replicate, and controls can be run simultaneously. If the apple juice is not filtered, then there can be random spikes due to bits of apple that can confound the readings. Also, care needs to be taken to clean the injector to avoid a sticky mess. When using single-cuvette spectrophotometers, we randomly assign each research team only three or four of the substrate concentrations and then assemble the class data. This method is easier for the students to trouble shoot. Both strategies work quite well.

The inhibitors, benzoic acid (122.12 g/mol) and phenythiourea (PTU) (152.04 g/mol), are not very water soluble, so we prepare them in 80% ethanol in water. This solvent should be added to controls without inhibitor (final ethanol concentration will be about 11.4%). It is easiest to make a single large batch of 80% ethanol to use for solution making and controls. We used stock solutions of 20 mM benzoic acid and 0.4 mM PTU (to avoid weighing impossibly small amounts, we prepare the latter in two steps) added so that final reaction concentrations are 2.86 mM benzoic acid (competitive) and .057 mM phenylthiourea (noncompetitive). Benzoic acid is a relatively safe compound, but students should avoid skin or eye exposure (acidic), whereas PTU is quite toxic, especially orally.

Routine safety precautions should be implemented. Gloves are recommended when handing the catechol, benzoic acid, and PTU solutions. The waste and left overs should be collected and disposed of properly, not poured down the sink.

Notes on data analysis

Catechol oxidase might be an easy enzyme to extract, but the data analysis can get tricky. The product, o-quinone, has to stack, forming aggregates, for the dark brown color to appear. Therefore at the lowest catechol concentrations, there may be a delay before a linear reaction rate is observed. At the highest catechol concentrations, if the enzyme activity is high, then there is likely to be an apparent slowing of the rate as initial conditions are no longer met and there may be an apparent decrease in product formation due to precipitation of large aggregates. This means that students must look carefully at the rate data and select the linear regions; they will not be in the same time frame for all catechol concentrations! Although students do not like this fussy step, which is hard to automate, it is a great example of what needs to be done to collect data that can be analyzed.

5.3.1 Notes on other enzymes or similar experiments

Predator–prey or the enzyme game

There are many versions of a "dry lab" exercise to generate Holling Type II or Michaeli–Menten-like data. In fact, the Holling "disc" equation comes from one such exercise in which discs of sandpaper "prey" were placed on the floor for a blindfolded subject (predator) to pick up and place in a dish. The prey density is varied by varying the number of discs in the arena, that is, the concentration. This can be done with sandpaper discs, beans, or dry macaroni scattered on the lab bench in a defined area (e.g., a square meter or less). Assistants measure the handling time and maintain the prey concentration by adding a new prey organism each time one is "captured."

To simulate an enzyme, the "prey" is now described as the substrate. The turnover number of the enzyme is controlled by a timer allowing, for example, only one attempt per second. The beans or other object are picked up and dropped in a cup to be counted at the end of a set time period. To keep the concentration constant, a bean can be added each time one is taken away. Competitive inhibition occurs if another type of bean is added to the mix.

For both simulations, replicates should be done at each of 6–8 concentrations or densities. For example, in a one meter by one meter arena, 4, 10, 30, 90, 180, 270 "prey" or substrate molecules work well.

Other enzymes

There are hundreds of enzymes that could be suitable for this lab, and your choice might be based on access or familiarity. The dehydrogenases are reasonably easy to use as the reaction can be observed continuously, and many, such as alcohol dehydrogenase (ADH) from yeast, can be purchased in pure form. ADH catalyzes the oxidation of ethanol to acetaldehyde by reducing NAD^+ to NADH, which can be detected by absorbance at 340 nm ($\epsilon = 6222$ $Lmol^{-1}$ cm^{-1}). The K_M is relatively high (16–19 mM), so that a reasonable substrate (ethanol) concentration range in the reaction mixture is from 0.002 M to 0.133 M prepared in distilled water with the concentration of NAD^+ fixed at 0.25 mM in the cuvette. A reasonable final ADH concentration is 0.17 µg/mL or about 1.18×10^{-9} M prepared from a stock of 1 µg/mL and diluted 1:5. The reaction works best at high pH, 0.1 M Tris-HCl at pH 8.5–9 in the reaction mixture. To keep the measurements easy, we prepare the NAD and ethanol stock solutions at 3 × their final values, the buffer, and enzyme stocks at 6 × their final values and combine these stocks in the ratio 2 NAD^+:2

ethanol:1 buffer mixed well and start the reaction with one volume of enzyme stock mixing well: the actual volumes depend on the size of the cuvette or well in a plate reader. Note that the absorbance of NADH is at 340 nm and best observed in acrylic disposable or quartz cuvettes vs. typical polystyrene disposable ones. At least one competitive inhibitor, 2,2,2-trifluoroethanol ($K_i = 2.85$ mM), can be used with good results at 1 to 3 mM [67]. A noncompetitive inhibitor quercitin works by removing one of the critical zinc atoms from the enzyme, thus rendering it nonfunctional: useful concentrations are 20–30 μM [6].

Other dehydrogenases include succinate dehydrogenase readily measured in mitochondrial isolations or lactate dehydrogenase, which is reliably obtained from crude muscle extracts.

5.4 Blood glucose monitoring

The goals of this lab are to collect data on one variable (in this case blood glucose) and use a model to predict the behavior of a second variable, insulin. The model is calibrated only using the blood glucose concentration dynamics data and several different ways to model its uptake rate from the intestines.

5.4.1 Tips for glucose monitoring

Due to the availability of easy and safe-to-use personal glucose monitoring supplies, this lab is actually technically fairly easy and safe to do, especially if the guidelines are followed [43]. We find it important to give students a week to consider their participation prior to signing the permission forms (an example is provided at the end of this section) and making it abundantly clear that students are not required to be the subject. Data are fairly reliable, so having as few as two or three subjects is adequate. We find however that most students truly want to be subjects, and so we spend more time emphasizing caution and reasons to consider participating as an assistant vs. taking their own blood samples. We also demonstrate proper interactions between subject and assistant, for example, actively asking permission to take the sample or hold the hand in place.

It is a good idea to consider what you will do if a student feels faint from the sight of blood or fasting too long. Remind students to stay hydrated. BandAids and soothing antibiotic cream are good to have on hand for fingers that have been overzealously poked.

We suggest touching base with the Human Subjects Review Board (IRB) at your institution to make sure that you are in compliance with your institutional policies and adjust the permission form to reflect any concerns that they may have.

Similarly, be sure you are complying with hazardous or biological waste in compliance for your institution. We treat the used glucose strips, alcohol wipes, and blood-stained diaper paper as "biological waste" and the put used lancets in sharps disposal containers.

If the data of a subject seem abnormal (e.g., glucose remains high, spikes higher than expected), then remember that this is not a diagnostic lab and that you are not a physician. Do encourage the student to ask about his/her glucose tolerance at a future medical appointment. Higher than expected initial concentrations of blood glucose are observed if the subject ate a meal prior to lab: this is normal.

Although students generally prefer to use their own data, you may wish to (with student permission) share particularly good data sets with the entire class. The glucose rise and fall will be complete in 40 to 60 min.

Blood glucose monitors and supplies are moderately expensive and can be purchased through the local pharmacy or through multiple online suppliers: some offer regular test-strip shipments. You will use more glucose test strips (and lancets and alcohol wipes) than you anticipate as students will take extra points or keep trying if a reading fails. Batteries for the glucose monitors are also good to have on-hand as they begin to be "fussy" as the battery gets low.

Follow up activities abound, and students enjoy choosing their own. Pasta or protein give interesting results. Tasting but not swallowing sweets can lower blood sugar. Exercise, if very intense, will also lower blood sugar. Caffiene and energy drinks have some confounding effects that make students think carefully.

Example of a subject consent form

Glucose Lab Consent Form

STUDY OVERVIEW. Subjects will come to lab after a 2 hour fast and then consume a source of sugar, such as juices and jelly beans, or other foods such as pure protein and/or perform various forms of exercise. Blood glucose will be measured initially (before consumption or other activities) and at 15–20 minute intervals over the two-hour lab period. A sterile lancet will be used by the subject (or a lab partner assistant) to draw a drop of blood. Standard glucose meters will measure plasma glucose concentrations over time. These procedures are routinely used by diabetic individuals and considered to be of low risk. Students will work in pairs, one who will be the subject ingesting glucose or exercising and taking a blood sample and another who will be recording the data or assisting the subject as desired. Afterwards, the students will analyze the data in the context of mathematical models for homeostatic control.

STUDY DETAILS. A commercial brand of lancet, chosen for its ease of use and safety record, will be used to obtain a small drop of blood from a finger tip. The sharp tips are protected during loading and by the lancet holder during use: they will be disposed immediately in a sharps container to avoid accidental pokes. We will use easy to operate glucometers. Test strips are loaded without any contact with the strips. Once the blood sample has been read, test strips are disposed with the push of a button into a biological waste container without any contact.

We can obtain excellent results if only 50% or less of the class participate as subjects. The rest of the students record data, and some may be asked to use the lancet on their partners and/or take the reading with the glucose monitor. Gloves must be worn by all assistants. The partner performing the testing MUST ask before every procedure and prior to any physical contact with the donor student.

WHO SHOULD NOT PARTICIPATE. If the following characteristics apply to you, then we please ask you not to participate as a subject: o diabetes o blood borne illness or infection o difficulty with sight of blood o unable or unwilling to eat or drink the sugar sources o unable or unwilling to perform the exercise activities o feeling lightheaded, dizzy, or weak o fasted for over 12 hours o propensity to faint o prefer not to have plasma glucose concentration made public o prefer not to be poked by a lancet multiple times.

Students who choose not to participate will not be questioned, punished, penalized, or in any way affected by that decision. Students who do choose to participate are free to limit or discontinue their participation at any time. Students who participate should know that their personal blood glucose concentration behavior as obtained during these experiments will be known by the students recording the values, your professors, and likely by others in the class. It should not be shared further. We will emphasize confidentiality.

RISKS TO HEALTHY INDIVIDUALS. Physical: Subjects will have blood drawn by the needle prick of a lancet, which will occur multiple times. This may hurt, especially after a few times. Students may become energized by the amount of sugar contained in each sugar source, but it is not considered an excessive amount, probably typical of something we do several times a day. Students who choose to exercise may face potential exhaustion: they may discontinue the studies at any time.

Emotional: Students will inevitably have some knowledge of their peer's glucose concentrations; this has the potential to be embarrassing should the values reveal information about the student's health. Students who feel uncomfortable about this possibility should not agree to participate. Regardless, these data are **not** to be shared outside of the class.

We consider these risks to be minimal. The process is a common one that has been perfected to operate smoothly and safely. Remember that a subject may discontinue participation at any time with no penalty.

If you have any questions or concerns, please feel free to bring them up confidentially with your instructor: INSTRUCTOR'S NAME HERE

Should you wish to participate, please fill out the following form.

I, (PRINT NAME HERE), agree to participate as a donor subject in the *Hormones and Homeostasis: Keeping Blood Glucose Concentrations Stable*. I understand the risks and concerns, I have read the details of the experiment, and I volunteer willingly as a subject in the experiments outlined above. I have no known risk factors that might put me or others in danger should I participate.

Signature:

5.4.2 Other lab activities

The Lotka–Volterra competition model or predator–prey model introduced in Unit 3 can be solved using the `deSolve` package in R, and parameters found using `modfit`.

Thus experimental systems in which two species compete or one serves as predator and the other as prey can make exciting labs or projects. Although they can be set up during a typical lab period, data collection rates will depend on the growth rates of species of interest and may extend over a period of a week or more. Live systems are tricky, and the best to use depends on the expertise of the biologists available to you. Many binary predator–prey models have a tendency to crash to extinction as they are too simple to be realistic. However, it should be possible to set up binary competition experiments in which two similar species compete for a single resource by comparing the growth rate of each species in a monoculture compared to when they are cocultured. For example, ciliates such as *Paramecium* may compete for the nutrients in their broth. A series of experiments using *Paramecia* species and other ciliates was developed by Jon C. Glase and is available through the Ecological Society of America. Another experimental system compares two species of parasitoid wasps and their reproductive success on their host, the pupae of *Neobellieria bullata* as described in another online resource from the teaching section of the Ecological Society of America. The resource also provides data sets to use in exercises of your choice. Plant growth competition experiments using peas and corn, for example, planting 8 pea, 8 corn, and combinations of corn and pea seeds in small containers and comparing the biomass after a period of several weeks should work but do take a few weeks to complete. Finally, if students enjoyed their bacterial or yeast growth labs, then it is possible to compare growth of Gram+ vs. Gram– bacteria (in separate and cocultures) using plate counts and Gram staining protocols. Similar experiments can be done with two species of yeast.

The dry lab described for Unit 3 to model functional responses can readily be modified to become a predator–prey lab and the data used to solve the Lotka–Volterra predator–prey model. There are many versions of this exercise, which is widely used at the K-12 and college level. One description is provided by Bridget Henshaw on Cornell Institute for Biology Teacher's website: https://blogs.cornell.edu/cibt/labs-activities/labs/predator-pray-population-oscillation/.

Bibliography

[1] Bovas Abraham, Johannes Ledolter, Introduction to Regression Modeling, Thomson Brooks-Cole, 2006.

[2] Matthew P. Adams, Catherine J. Collier, Sven Uthicke, Yan X. Ow, Lucas Langlois, Katherine R. O'Brien, Model fit versus biological relevance: evaluating photosynthesis-temperature models for three tropical seagrass species, Scientific Reports 7 (2017) 39930.

[3] Vanndita Bahl, Parth Kapoor, Prasidhi Tyagi, Y.V.R. Kameshwar Sharma, Enzymatic determination of catechol oxidase and protease from fruits (orange, apple) and vegetables (carrot, tomato), IOSR Journal of Pharmacy and Biological Sciences 5 (5) (2013) 29–35.

[4] Sébastien Benzekry, Clare Lamont, Afshin Beheshti, Amanda Tracz, John M.L. Ebos, Lynn Hlatky, Philip Hahnfeldt, Classical mathematical models for description and prediction of experimental tumor growth, PLoS Computational Biology 10 (8) (2014) e1003800.

[5] Craig Bettenhausen, Engineered apples near approval: fruit with nonbrowning genes may get green light in U.S., Chemical & Engineering News 91 (2013) 31–33.

[6] Sutanwi Bhuiya, Lucy Haque, Ankur Bikash Pradhan, Suman Das, Inhibitory effects of the dietary flavonoid quesrcetin on the enzyme activity of zinc(ii)-dependent yeast alcohol dehydrogenase: spectroscopic and molecular docking studies, International Journal of Biological Macromolecules.

[7] N. Broekhuizen, M.P. Hassell, H.F. Evans, Common mechanisms underlying contrasting dynamics in two populations of the pine looper moth, Journal of Animal Ecology 63 (2) (1994) 245–255.

[8] Matthew J. Butler, Grant Harris, Bradley N. Strobel, Influence of whooping crane population dynamics on its recovery and management, Biological Conservation 162 (2013) 89–99.

[9] Matthew J. Butler, Kristine L. Metzger, Grant Harris, Whooping crane demographic responses to winter drought focus conservation strategies, Biological Conservation 179 (2014) 72–85.

[10] Jaime E. Cascante, Maria C. Rojas, Armando Salinas, Mariajosé Serna, Henry D. Torres, Esteban Vargas, Juan M. Cordovéz, Jose R. Arteaga, Influence of brood deaths on honey bee population dynamics and the potential impact of insecticides, International Research Experience for Students, https://mcmsc.asu.edu/ires, 2017.

[11] M. Deanna Colton, Eric V. Stabb, Stephen J. Hagen, Modeling analysis of signal sensitivity and specificity by *Vibrio fischeri* luxr variants, PLoS ONE 10 (5) (2015) e0126474.

[12] Peter Convey, Reproduction of Antarctic flowering plants, Antarctic Science 8 (2) (1996) 127–134.

[13] Curtis A. Deutsch, Joshua J. Tewksbury, Raymond B. Huey, Kimberly S. Sheldon, Cameron K. Ghalambor, David C. Haak, Paul R. Martin, Impacts of climate warming on terrestrial ectotherms across latitude, Proceedings of the National Academy of Sciences 105 (18) (2008) 6668–6672.

[14] Charles G. Drake, Prostate cancer as a model for tumour immunotherapy, Nature Reviews Immunology 10 (2010) 580–593.

[15] Leah Edelstein-Keshet, Mathematical Models in Biology, vol. 46, Siam, 1988.

[16] James J. Elser, Tom Andersen, Jill S. Baron, Ann-Kristin Bergström, Mats Jansson, Marcia Kyle, Koren R. Nydick, Laura Steger, Dag O. Hessen, Shifts in lake N:P stoichiometry and nutrient limitation driven by atmospheric nitrogen deposition, Science 326 (5954) (2009) 835–837.

[17] Jeffrey A. Evans, Adam S. Davis, Consequences of parameterization and structure of applied demographic models: a comment on Pardini et al. (2009), Ecological Applications 21 (2) (2011) 608–613.

[18] Agnes Fekete, Christina Kuttler, Michael Rothballer, Burkhard A. Hense, Doreen Fischer, Katharina Buddrus-Schiemann, Marianna Lucio, Johannes Müller, Philippe Schmitt-Kopplin, Anton Hartmann, Dynamic regulation of N-acyl-homoserine lactone production and degradation in *Pseudomonas putida* IsoF, FEMS Microbiology, Ecology 72 (1) (2010) 22–34.

[19] Nicholas Gruber, James N. Galloway, An Earth-system perspective of the global nitrogen cycle, Nature 451 (2008) 293–296.

[20] Shailendra K. Gupta, Tanushree Jaitly, Ulf Schmitz, Gerold Schuler, Olaf Wolkenhauer, Julio Vera, Personalized cancer immunotherapy using systems medicine approaches, Briefings in Bioinformatics 17 (2016) 453–467.

[21] James W. Haefner, Modeling Biological Systems: Principles and Applications, Springer Science & Business Media, 2005.

[22] Ernst Hairer, Syvert P. Nørsett, Gerhard Wanner, Solving Ordinary Differential Equations I, Springer Series in Computational Mathematics, vol. 8, Springer-Verlag, Berlin, 1993.

[23] Ernst Hairer, Gerhard Wanner, Solving Ordinary Differential Equations II: Stiff and Differential-Algebraic Problems, Springer Series in Computational Mathematics, vol. 14, 1996.

[24] Harris James Lloyd, A population model and its application to the study of honey bee colonies, 1980.

[25] Jean-Claude Henquin, Denis Dufrane, Myriam Nenquin, Nutrient control of insulin secretion in isolated normal human islets, Diabetes 55 (12) (2006) 3470–3477.

[26] J. Iacarella, J.A. Dick, M.E. Alexander, A. Ricciardi, Ecological impacts of invasive alien species along temperature gradients: testing the role of environmental matching, Ecological Applications 25 (2015) 706–716.

[27] Meyer B. Jackson, Molecular and Cellular Biophysics, Cambridge University Press, 2006.

[28] Steven Johnson, The Ghost Map: The Story of London's Most Terrifying Epidemic–and How It Changed Science, Cities, and the Modern World, Penguin, 2006.

[29] S. Juliano, Nonlinear curve fitting: predation and functional response curves, in: Nonlinear Curve Fitting. Design and Analysis of Ecological Experiments, Chapman and Hall, New York, New York, USA, 2001.

[30] Andries Kalsbeek, Susanne la Fleur, Eric Fliers, Circadian control of glucose metabolism, Molecular Metabolism 3 (2014) 372–383.

[31] Santosh R. Kanade, V.L. Suhas, N. Chandra, Lalitha R. Gowda, Functional interaction of diphenols with polyphenol oxidase: molecular determinants of substrate/inhibitor specificity, The FEBS Journal 274 (2007) 4177–4187.

[32] Raphael Barroso Kato, Victor Srougi, Fernanda Aburesi Salvadori, Pedro Paulo Marino Rodrigues Ayres, Katia Moreira Leite, Miguel Srougi, Pretreatment tumor volume estimation based on total serum psa in patients with localized prostate cancer, Clinics 63 (6) (2008) 759–762.

[33] Denise Kirschner, John Carl Panetta, Modeling immunotherapy of the tumor–immune interaction, Journal of Mathematical Biology 37 (3) (1998) 235–252.

[34] Thomas Klabunde, Christoph Eicken, James C. Sacchettini, Bernt Krebs, Crystal structure of a plant catechol oxidase containing a dicopper center, Nature Structural Biology 5 (12) (1988) 1084–1090.

[35] Natalie Kronik, Yuri Kogan, Moran Elishmereni, Karin Halevi-Tobias, Stanimir Vuk-Pavlović, Zvia Agur, Predicting outcomes of prostate cancer immunotherapy by personalized mathematical models, PLoS ONE 5 (12) (2010) e15482.

[36] S. Casey Laizure, Bernd Meibohm, Kembral Nelson, Feng Chen, Zhe-Yi Hu, Robert B. Parker, Comparison of caffeine disposition following administration by oral solution (energy drink) and inspired powder (aeroshot) in human subjects, British Journal of Clinical Pharmacology 83 (12) (2017) 2687–2694.

[37] Fang Liu, Jin-Hong Zhao, Zhi-Lin Gan, Yuan-Ying Ni, Comparison of membrane-bound and soluble polyphenol oxidase in Fuji apple (*Malus domestica* borkh. cv. red Fuji), Food Chemistry 173 (2015) 86–91.

[38] Jonathan B. Losos, Robert E. Ricklefs, The Theory of Island Biogeography Revisited, Princeton University Press, 2009.

[39] Donald Ludwig, Dixon D. Jones, Crawford S. Holling, Qualitative analysis of insect outbreak systems: the spruce budworm and forest, The Journal of Animal Ecology 47.

[40] Robert H. MacArthur, Edward O. Wilson, An equilibrium theory of insular zoogeography, Evolution 17 (4) (1963) 373–387.

[41] Robert H. MacArthur, Edward O. Wilson, The Theory of Island Biogeography, Princeton University Press, 1967.

[42] William Mendenhall, Terry Sincich, Nancy S. Boudreau, A Second Course in Statistics: Regression Analysis, vol. 5, Prentice Hall, Upper Saddle River, NJ, 1996.

[43] Robert K. Moats, Blood glucose monitoring as a teaching tool for endocrinology: a new perspective, Advances in Physiological Education 33 (2009) 209–212.

[44] Lars Möhler, Dietrich Flockerzi, Heiner Sann, Udo Reichl, Mathematical model of influenza a virus production in large-scale microcarrier culture, Biotechnology and Bioengineering 90 (1) (2005) 46–58.

[45] Kenneth A. Nagy, Isabelle A. Girard, Tracey K. Brown, Energetics of free-ranging mammals, reptiles, and birds, Annual Review of Nutrition 19 (1) (1999) 247–277.

[46] Van Kinh Nguyen, Sebastian C. Binder, Alessandro Boianelli, Michael Meyer-Hermann, Esteban A. Hernandez-Vargas, Ebola virus infection modeling and identifiability problems, Frontiers in Microbiology 6 (2015) 257.

[47] Lulla Opatowski, Christophe Fraser, Jamie Griffin, Eric De Silva, Maria D. Van Kerkhove, Emily J. Lyons, Simon Cauchemez, Neil M. Ferguson, Transmission characteristics of the 2009 H1N1 influenza pandemic: comparison of 8 Southern hemisphere countries, PLoS Pathogens 7 (9) (2011) e1002225.

[48] Eleanor A. Pardini, John M. Drake, Jonathan M. Chase, Tiffany M. Knight, Complex population dynamics and control of the invasive biennial *Alliaria petiolata* (garlic mustard), Ecological Applications 19 (2) (2009) 387–397.

[49] Eleanor A. Pardini, John M. Drake, Tiffany M. Knight, On the utility of population models for invasive plant management: response to Evans and Davis, Ecological Applications 21 (2) (2011) 614–618.

[50] Huiming Peng, Weiling Zhao, Hua Tan, Zhiwei Ji, Jingsong Li, King Li, Xiaobo Zhou, Prediction of treatment efficacy for prostate cancer using a mathematical model, Nature: Scientific Reports 6 (2016) 21599.

[51] M.R. Penk, J.M. Jeschke, D. Minchin, I. Donahue, Warming can enhance invasion success through asymmetries in energetic performance, Journal of Animal Ecology 85 (2016) 419–426.

[52] Pablo Delfino Pérez, Stephen J. Hagen, Heterogeneous response to a quorum-sensing signal in the luminescence of individual *Vibrio fischeri*, PLoS ONE 5 (11) (2010) e15473.

[53] Leon Perrie, Patrick Brownsey, Molecular evidence for long-distance dispersal in the New Zealand pteridophyte flora, Journal of Biogeography 34 (12) (2007) 2028–2038.

[54] Daniel W. Pritchard, Rachel Paterson, Helene C. Bovy, Daniel Barrios-O'Neill, Frair: an R package for fitting and comparing consumer functional responses, Methods in Ecology and Evolution 8 (11) (2017) 1528–1534.

[55] Song S. Qian, Environmental and Ecological Statistics with R, CRC Press, 2016.

[56] Navneet Rai, Rajat Anand, Krishna Ramkumar, Varun Sreenivasan, Sugat Dabholkar, K.V. Venkatesh, Mukund Thattai, Prediction by promoter logic in bacterial quorum sensing, PLoS Computational Biology 8 (1) (2012) e1002361.

[57] Emma Louise Rawlins, Brigid L.M. Hogan, Ciliated epithelial cell lifespan in the mouse trachea and lung, American Journal of Physiology. Lung Cellular and Molecular Physiology 295 (2008) L231–L235.

[58] R.E. Ricklefs, E. Bermingham, Taxon cycles in the Lesser Antillean avifauna, Ostrich 70 (1) (1999) 49–59.

[59] A.M.C.N. Rocha, A.M.M.B. Morais, Characterization of polyphenoloxidase (PPO) extracted from 'Jonagored' apple, Food Control 12 (2001) 85–90.

[60] Lars Gösta Rudstam, Amy Lee Hetherington, Ali Martonius Mohammadian, Effect of temperature on feeding and survival of *Mysis relicta*, Journal of Great Lakes Research 25 (2) (1999) 363–371.

[61] R. Sanft, A. Walter, Experimenting with mathematical biology, PRIMUS 26 (1) (2015) 83–103.

[62] Irwin H. Segel, Biochemical Calculations: How to Solve Mathematical Problems in General Biochemistry, 2nd edition, John Wiley and Sons, 1976.

[63] Lauralee Sherwood, Hillar Klandorf, Paul H. Yancey, Animal Physiology From Genes to Organisms, second edition, Brooks/Cole, 2013.

[64] J.S. Singh, S.R. Gupta, Plant decomposition and soil respiration in terrestrial ecosystems, The Botanical Review 43 (4) (1977) 449–528.

[65] Karline Soetaert, Peter M.J. Herman, A Practical Guide to Ecological Modeling: Using R as a Simulation Platform, Springer Science & Business Media, 2008.

[66] K.E.R. Soetaert, Thomas Petzoldt, R. Woodrow Setzer, Solving differential equations in R: package desolve, Journal of Statistical Software 33 (2010).

[67] Richard L. Taber, The competitive inhibition of yeast alcohol dehydrogenase by 2, 2, 2-trifluoroethanol, Biochemical Education 26 (3) (1998) 239–242.

[68] John Terborgh, Chance, habitat and dispersal in the distribution of birds in the West Indies, Evolution 27 (2) (1973) 338–349.

[69] I.W.B. Thornton, R.A. Zann, S. Van Balen, Colonization of Rakata (Krakatau Is.) by non-migrant land birds from 1883 to 1992 and implications for the value of island equilibrium theory, Journal of Biogeography (1993) 441–452.

[70] Eric J. Ward, A review and comparison of four commonly used Bayesian and maximum likelihood model selection tools, Ecological Modelling 211 (1) (2008) 1–10.

[71] John Ward, Mathematical modeling of quorum-sensing control in biofilms, in: Control of Biofilm Infections by Signal Manipulation, Springer, 2008, pp. 79–108.

[72] R.J. Whittaker, M.B. Bush, K.J.E.M. Richards, Plant recolonization and vegetation succession on the Krakatau Islands, Indonesia, Ecological Monographs 59 (2) (1989) 59–123.

[73] Robert J. Whittaker, José María Fernández-Palacios, Thomas J. Matthews, Michael K. Borregaard, Kostas A. Triantis, Island biogeography: taking the long view of nature's laboratories, Science 357 (6354) (2017) eaam8326.

[74] Kyle A. Whittinghill, William S. Currie, Donald R. Zak, Andrew J. Burton, Kurt S. Pregitzer, Anthropogenic N deposition increases soil C storage by decreasing the extent of litter decay: analysis of field observations with an ecosystem model, Ecosystems 15 (3) (2012) 450–461.

[75] Jianguo Wu, John L. Vankat, Island biogeography: theory and applications, Encyclopedia of Environmental Biology 2 (1995) 371–379.

[76] X.S. Xie, Enzyme kinetics, past, present and future, Science 342 (2013) 1457–1459.

[77] Weikai Yan, L.A. Hunt, An equation for modelling the temperature response of plants using only the cardinal temperatures, Annals of Botany 84 (5) (1999) 607–614.

[78] Zhike Zi, Sensitivity analysis approaches applied to systems biology models, IET Systems Biology 5 (6) (2011) 336–346.

Index

E

F

G

H

I

L

R Quick Help

Getting help

`?plot`
Access documentation for a function

Libraries

`install.packages("deSolve")`
Downloads package from CRAN and installs package.

`library(deSolve)`
Loads package into session.

For loops

```
y = rep(0,11)
y[1] = 2

for (n in 1:(length(y)-1)){
  y[n+1] = 1.1*y[n]
}
```

Iterates equation $y_{n+1}=1.1y_n$ and stores results in a vector y.
Examples: Listings 2.1, 2.3, 2.7, 2.8, 3.5, 4.15, 4.17, 4.18

If statements

```
if (t<10){
  g = 50 }
else {
  g = 0 }
```

If expression (t<10) is TRUE, the following statement (g=50) is executed. If expression is FALSE, g=0 is executed.
Example: Exercise 4.51

Math functions

`log(x)`	natural log	`max(x)`	largest element
`exp(x)`	exponential	`min(x)`	smallest element
`sin(x)`	sine	`sum(x)`	sum of elements
`cos(x)`	cosine	`mean(x)`	mean of elements

Vectors

Generating Vectors	
`c(1,3,4)`	Joins elements into a vector: 1 3 4
`2:5`	Sequence: 2 3 4 5
`seq(1,2,by=0.2)`	Sequence: 1 1.2 1.4 1.6 1.8 2
`rep(0,4)`	Repeat a number: 0 0 0 0

Selecting Vector Elements	
`v[2]`	Second element of v
`v[2:5]`	Elements 2 through 5
`v[-2]`	All but the second element of v
`which(v %in% c(2,5))`	Indices of v that correspond to elements equal to 2 or 5

Matrices

```
M = matrix(x, nrow = 3,
ncol = 2, byrow = FALSE)
```

Creates matrix with 3 rows and 2 columns from vector x, matrix is filled by columns if byrow = FALSE.

Selecting Matrix Elements	
`M[2,1]`	Element of M in second row, first column
`M[2,]`	Second row of M
`M[,2]`	Second column of M

Data frames

`df = data.frame(t=0:2,y=c(2,5,9))`
Creates data frame with first column named t and second column named y.

`df$t`
Returns the t column as a vector.

`nrow(df)`
Returns number of rows in `df`.

`head(df)`
Returns first six rows of `df`.

`as.data.frame(M)`
Converts matrix M to a data frame.

`rbind(df, c(3,14))`
Adds vector c(3,14) to last row of `df`.

`ncol(df)`
Returns number of columns in `df`.

`tail(df)`
Returns last six rows of `df`.

Plotting

```
plot(x, y, type = "p", main = "title",
      xlab = "x axis", ylab = "y axis", col = "red",
        xlim = c(x1,x2), ylim = c(y1,y2), pch = 20)
```

Plots y versus x, type="p" for points, type="l" for lines, type="b" for both.

```
lines(x,y)
```

Adds line graph to plot.

```
points(x,y)
```

Adds points to a plot.

```
legend("topleft", legend=c("a", "b"), lty=c(1,NA),
pch=c(NA,20), col=c("black", "blue"))
```

Includes a legend located in top left of graph (can also use coordinates for positioning). Here the first curve is a black (col="black") solid line (lty=1) with no point markers (pch=NA) called "a". The second curve is a blue (col="blue") scatter plot (pch=20) called "b".

```
plot(NULL, xlim = c(0,1), ylim = c(0,10))
```

Draws an empty plot.

```
barplot(x, main = "title", names.arg=c("a", "b", "c"))
```

Plots a bar plot with heights given by the values in the vector x. names.arg is a vector of names to be plotted below each bar. Example: Listing 4.12

```
plot_ly(x,y,z,type="heatmap")
```

Creates a heatmap where x is a vector, y is a vector, and z is a matrix with values that correspond to every combination of pairs of elements from x and y. The library plotly must be installed and loaded. Examples: Listings 3.5 and 4.15

Sliders

```
manipulate({
  t = seq(0,5,by=0.1)
  plot(t, y0*exp(r*t), type="l", ylim=c(0,50))},
  r = slider(0,0.5,step=0.01,initial=0.1),
  y0 = slider(0,5,step=0.1, initial=1))
```

Creates plot with sliders to dynamically change parameters. The library manipulate must be installed and loaded.
Examples: Listings 3.1, 3.6, and 4.7

Fitting Models to Data

```
linearMod = lm(y~x)
```

Fits a linear model where y is the dependent variable and x is the independent variable. Results assigned to linearMod. `lm(y~x+0)` forces line to go through origin. Examples: Listings 2.2, 2.5, and 2.6

```
summary(linearMod)
```

Prints summary statistics.

```
abline(linearMod)
```

Adds regression line to existing plot.

```
nonlinMod = nls(y~a*exp(-b*x), start=list(a=0.1, b=10))
```

Determines nonlinear least-squares estimates of parameters of a nonlinear model, where y is the dependent variable and x is the independent variable. The function (ae^{-bx} in this example) and starting values for the unknown parameters must be specified. Examples: Listings 3.2, 3.7, and 3.8

```
resid(nonlinMod)
```

Prints model residuals.

```
predict(nonlinMod, list(x = xvals))
```

Evaluates model at values given in vector xvals.

```
library(FME)
modFit(f, p, lower=c(...), upper=c(...))
```

Fits a model to data. The function `f` must be defined and must return a vector of residuals (between model and data). Input `f` is a vector of parameters to be optimized over, and lower and upper bounds for parameters can be included. Examples: Listings 4.6 and 4.8

Solving Differential Equations

```
# Function evaluates right side of differential equations
diffeq = function(t,y,parms) {
   with(as.list(c(y,parms)),{
        dv = r1*v - a*v*p
        dp = b*v*p - r2*p
        list(c(dv,dp) })}

out = ode(y0 = c(v=4,p=2), times = seq(0,20,by=0.01),
              diffeq, parms = c(r1=2, r2=1, a=1.2, b=0.9))
```

Solves system of differential equations given in function called diffeq. Inputs to function ode include initial conditions, a time vector, the function where system is defined, and parameters. Library deSolve must be installed and loaded.
Examples: Listings 4.1, 4.2, 4.3, 4.4, 4.14, and 4.16

Printed in the United States
By Bookmasters